Balanced Assessment for the Mathematics Curriculum

Elementary Grades

BERKELEY
HARVARD
MICHIGAN STATE
SHELL CENTRE

Balanced Assessment for the
Mathematics Curriculum

Dale Seymour Publications®

Project Directors: Alan Schoenfeld
Hugh Burkhardt
Phil Daro
Jim Ridgway
Judah Schwartz
Sandra Wilcox

Managing Editors: Catherine Anderson and Alan MacDonell

Acquisitions Editor: Merle Silverman

Project Editor: Toni-Ann Guadagnoli

Production/Manufacturing Director: Janet Yearian

Senior Production/Manufacturing Coordinator: Fiona Santoianni

Design Director: Phyllis Aycock

Design Manager: Jeff Kelly

Text and Interior Designer: Don Taka

Cover Image: Hutchings Photography

Illustrations: Larry Nolte

The work of this project was supported by a grant from the National Science Foundation. The opinions expressed in these materials do not necessarily represent the position, policy, or endorsement of the Foundation.

Copyright © 1999 by the Regents of the University of California. All rights reserved.

This book is published by Dale Seymour Publications®,
an imprint of Addison Wesley Longman, Inc.

Dale Seymour Publications
10 Bank Street
White Plains, NY 10602-5026
Customer Service: 800-872-1100

Limited reproduction permission: The publisher grants permission to individual teachers who have purchased this book to reproduce the blackline masters as needed for use with their own students. Reproduction for an entire school or school district or for commercial use is prohibited.

Printed in the United States of America
Order number 33001
ISBN 0-7690-0064-9

1 2 3 4 5 6 7 8 9 10-ML-02-01-00-99-98

This Book Is Printed
On Recycled Paper

Authors

This assessment package was designed and developed by members of the Balanced Assessment Project team, particularly Harold Asturias, Pam Beck, Rita Crust, Mishaa DeGraw, David Ott, and Richard Phillips. The editor was Mishaa DeGraw.

Many others have made helpful comments and suggestions in the course of the development. We thank them all. The Project is particularly grateful to the mathematics consultants, teachers, and students with whom these tasks were developed and tested, particularly Angela Archie, Donna Goldenstein, Linda Honeyman, Sandra Short, Verlinda Bailey, Lorrie Hugenburger, Julie Barkley, Kathy Freeburg, Bev Schmidt, Cheryl Prophet, Catherine Knowland, Susan Tognolini, Susie Leal, Janice Blumenkranz, Roseann Santa Cruz, Theresa Wong, Amy Hafter, Rosita Fabian, Laura Witson, Carolyn Granberg, Elisa Cotcher, Mollie Atkinson, Phil Tabano, and Michael Hunter.

The project was directed by Alan Schoenfeld, Hugh Burkhardt, Phil Daro, Sandra Wilcox, Judah Schwartz, and Jim Ridgway.

The package consists of materials compiled or adapted from work done at the four sites of the Balanced Assessment Project:

Balanced Assessment
Graduate School of Education
University of California
Berkeley, CA 94720-1670
USA

Balanced Assessment (MARS)
513 Erickson Hall
Michigan State University
East Lansing, MI 48824
USA

Balanced Assessment
Educational Technology Center
Harvard University
Cambridge, MA 02138
USA

Balanced Assessment
Shell Centre for Mathematical Education
University of Nottingham
Nottingham NG7 2RD
England

Additional tasks and packages, the materials in their original form, and other assessment resources such as guides to scoring may be obtained from the project sites. For a full list of available publications, and for further information, contact the Project's Mathematics Assessment Resource Service (MARS) at the Michigan State address above. We welcome your comments.

Table of Contents

What is balanced assessment?

Mathematics assessments tell us and our students how well they are learning mathematics. A carefully designed mathematics assessment should:

- assess the mathematics that counts, focusing on important ideas and processes;
- be fair to the students, providing them with a set of opportunities to demonstrate what they know and can do;
- be fair to the curriculum, offering a balance of opportunities—long and short tasks, basic knowledge and problem solving, individual and group work, and the spectrum of concepts and processes that reflect the vision of the NCTM *Standards;*
- be of such high quality that students and teachers learn from them—so that assessment time serves as instructional time, and assessment and curriculum live in harmony;
- provide useful information to administrators, so they can judge the effectiveness of their programs; to teachers, so they can judge the quality of their instruction; and to students and parents, so they can see where the students are doing well and where more work is needed.

This is such an assessment package, dealing with the mathematics appropriate for the elementary grades. It was designed by the Balanced Assessment Project, an NSF-supported collaboration that was funded to create a series of exemplary assessment items and packages for assessing students' mathematical performance at various grade levels (elementary grades, middle grades, high school, and advanced high school). Balanced Assessment offers a wide range of extensively field-tested tasks and packages—some paper-and-pencil, some high-tech or multimedia—and consulting services to help states and districts implement meaningful and informative mathematics assessments.

What is balance?

It's easy to see what isn't balanced. An assessment that focuses on computation only is out of balance. So is one that focuses on patterns, functions, and algebra to the exclusion of geometry, shape, and space, or that ignores or gives a cursory nod toward statistics and probability. Likewise, assessments that do not provide students with ample opportunity to show how they can reason or communicate mathematically are unbalanced. These are content and process dimensions of balance, but there are many others—length of task, whether tasks are pure or applied, and so on. The following table shows some of the dimensions used to design and balance this package. (For explanations of terms that may be unfamiliar, see the Glossary.)

Dimensions of Balance

Mathematical Content Dimension

- **Mathematical Content** will include some of the following:

  Number and Quantity including: concepts and representation; computation; estimation and measurement; number theory and general number properties.

  Patterns, Functions, and Algebra including: patterns and generalization; functional relationships (including ratio and proportion); graphical and tabular representation; symbolic representation; forming and solving relationships.

  Geometry, Shape, and Space including: shape, properties of shapes, relationships; spatial representation, visualization, and construction; location and movement; transformation and symmetry; trigonometry.

  Handling Data, Statistics, and Probability including: collecting, representing, and interpreting data; probability models—experimental and theoretical; simulation.

  Other Mathematics including: discrete mathematics, including combinatorics; underpinnings of calculus; mathematical structures.

Mathematical Process Dimension

- **Phases** of problem solving, reasoning, and communication will include, as broad categories, some or all of the following: modeling and formulating; transforming and manipulating; inferring and drawing conclusions; checking and evaluating; reporting.

Task Type Dimensions

- **Task Type** will be one of the following: open investigation; nonroutine problem; design; plan; evaluation and recommendation; review and critique; re-presentation of information; technical exercise; definition of concepts.
- **Nonroutineness** in: context; mathematical aspects or results; mathematical connections.
- **Openness:** It may have an open end with open questions; open middle.
- **Type of Goal** is one of the following: pure mathematics; illustrative application of the mathematics; applied power over the practical situation.
- **Reasoning Length** is the expected time for the longest section of the task. (It is an indication of the amount of "scaffolding"—the detailed step-by-step guidance that the prompt may provide.)

Circumstances of Performance Dimensions

- **Task Length:** ranging from short tasks (5–20 minutes), through long tasks (20–45 minutes), to extended tasks (several days to several weeks).
- **Modes of Presentation:** written; oral; video; computer.
- **Modes of Working** on the task: individual; group; mixed.
- **Modes of Response** by the student: written; built; spoken; programmed; performed.

What's in a package?

A typical Balanced Assessment Package offers ten to twenty tasks, ranging in length from 5 to 45 minutes. Some of the tasks consist of a single problem, while others consist of a sequence of problems. Taken together, the tasks provide students with an opportunity to display their knowledge and skills across the broad spectrum of content and processes described in the NCTM *Standards*. It takes time to get this kind of rich information—but the problems are mathematically rich and well worth the time spent on them.

What's included with each task?

We have tried to provide you with as much information as possible about the mathematics central to solving a task, about managing the assessment, and about typical student responses and how to analyze the mathematics in them. Each section of this package, corresponding to one task, consists of the following:

Overview The first page of each section provides a quick overview that lets you see whether the task is appropriate for use at any particular point in the curriculum. This overview includes the following:

- Task Description—the situation that students will be asked to investigate or solve.
- Assumed Mathematical Background—the kinds of previous experiences students will need to have had to engage the task productively.
- Core Elements of Performance—the mathematical ideas and processes that will be central to the task.
- Circumstances—the estimated time for students to work on the task; the special materials that the task will require; whether students will work individually, in pairs, or in small groups; and any other such information.

Task Prompt These pages are intended for the student. To make them easy to find, they have been designed with stars in the margin and a white bar across the top. The task prompt begins with a statement for the student characterizing the aims of the task. In some cases there is a pre-assessment activity that teachers assign in advance of the formal assessment. In some cases there is a launch activity that familiarizes students with the context but is not part of the formal assessment.

A Sample Solution Each task is accompanied by at least one solution; where there are multiple approaches to a problem, more than one may appear.

Using this Task Here we provide suggestions about launching the task and helping students understand the context of the problem. Some tasks have pre-activities; some have students do some initial exploration in pairs or as a whole class to become familiar with the context while the formal assessment is done individually. Information from field-testing about aspects of tasks that students may find challenging is given here. We may also include suggestions for subsequent classroom instruction related to the task, as well as extensions that can be used for assessment or instructional purposes.

Characterizing Performance This section contains descriptions of characteristic student responses that the task is likely to elicit. These descriptions, based on the *Core Elements of Performance,* indicate various levels of successful engagement with the task. They are accompanied by annotated artists' renderings of typical student work. These illustrations will prepare you to assess the wide range of responses produced by your students. We have chosen examples that show something of the range and variety of responses to the task, and the various aspects of mathematical performance it calls for. The commentary is intended to exemplify these key aspects of performance at various levels across several domains. Teachers and others have found both the examples and the commentary extremely useful; its purpose is to bring out explicitly for each task the wide range of aspects of mathematical performance that the standards imply.

Scoring student work

The discussions of student work in the section *Characterizing Performance* are deliberately qualitative and holistic, avoiding too much detail. They are designed to focus on the mathematical ideas that "count," summarized in the *Core Elements of Performance* for each task. They offer a guide to help teachers and students look in some depth at a student's work in the course of instruction, considering how it might be improved.

For some other purposes, we need more. Formal assessment, particularly if the results are used for life-critical decisions, demands more accurate scoring, applied consistently across different scorers. This needs more precise rubrics, linked to a clear scheme for reporting on performance. These can be in a variety of styles, each of which has different strengths. The Balanced Assessment Project has developed resources that support a range of styles.

For example, *holistic approaches* require the scorer to take a balanced overall view of the student's response, relating general criteria of quality in performance to the specific item. *Point scoring approaches* draw attention in detail to the various aspects of performance that the task involves, provide a natural mechanism for balancing greater strength in one aspect with some weakness in another, and are useful for *aggregating scores.*

How to use this package

This assessment package may be used in a variety of ways, depending on your local needs and circumstances.

- You may want to implement formal performance assessment under controlled conditions at the school, district, or state level. This package provides a balanced set of tasks appropriate for such on-demand, high-stakes assessment.
- You may want to provide opportunities for classroom-based performance assessment, embedded within the curriculum, under less-controlled conditions. This package allows you the discretion of selecting tasks that are appropriate for use at particular points in the curriculum.
- You may be looking for tasks to serve as a transition toward a curriculum as envisioned in the NCTM *Standards* or as enrichment for existing curriculum. In this case, the tasks in this package can serve as rich instructional problems to enhance your curriculum. They are exemplars of the kinds of instructional tasks that will support performance assessment and can be used for preparing students for future performance assessment. Even in these situations, the tasks provide you with rich sites to engage in informal assessment of student understanding.

Preparing for the assessment

We urge you to work through a task yourself before giving it to your students. This gives you an opportunity to become familiar with the context and the mathematical demands of the task, and to anticipate what might need to be highlighted in launching the task.

It is important to have at hand all the necessary materials students need to engage a task before launching them on the task. We assume that students have certain tools and materials available at all times in the mathematics classroom and that these will be accessible to students to choose from during any assessment activity.

At the elementary grades these resources include: calculators; rulers, standard and metric; grid paper, square and isometric dot paper; dice, square tiles, cubes, base 10 blocks, other concrete materials; scissors, markers, tape, string, paper clips, and glue.

If a task requires any special materials, these are specified in the task.

Managing the assessment

We anticipate that this package will be used in a variety of situations. Therefore, our guidance about managing assessment is couched in fairly general suggestions. We point out some considerations you may want to take into account under various circumstances.

The way in which any particular task is introduced to students will vary. The launch will be shaped by a number of considerations (for example, the students, the complexity of the instructions, the degree of familiarity students have with the context of the problem). In some cases it will be necessary only to distribute the task to students and then let them read and work through the task. Other situations may call for you to read the task to the class to assure that everyone understands the instructions, the context, and the aim of the assessment. Decisions of this kind will be influenced by the ages of the students, their experiences with reading mathematical tasks, their fluency with English, and whether difficulties in reading would exclude them from otherwise productively engaging with the mathematics of the task.

Under conditions of formal assessment, once students have been set to work on a task, you should not intervene except where specified. This is essential in formal, high-stakes assessment but it is important under any assessment circumstance. Even the slightest intervention—reinterpreting instructions, suggesting ways to begin, offering prompts when students appear to be stuck—has the potential to alter the task for the student significantly. However, you should provide general encouragement within a supportive classroom environment as a normal part of doing mathematics in school. This includes reminding students about the aim of the assessment (using the words at the beginning of the task prompt), when the period of assessment is nearing an end, and how to turn in their work when they have completed the task.

We suggest a far more relaxed use of the package when students are meeting these kinds of tasks for the first time, particularly in situations where they are being used primarily as learning tasks to enhance the curriculum. Under these circumstances you may reasonably decide to do some coaching, talk with students as they work on a task, or pose questions when they seem to get stuck. In these instances you may be using the tasks for informal assessment—observing what strategies students favor, what kinds of questions they ask, what they seem to understand and what they are struggling with, and what kinds of prompts get them unstuck. This can be extremely useful information in helping the you make ongoing instructional and assessment decisions. However, as students have more experiences with these kinds of tasks, the amount of coaching you do should decline and students should rely less on this kind of assistance.

Under conditions of formal assessment, you will need to make decisions about how tasks will be scored and by whom, how scores will be aggregated across tasks, and how students' accomplishments will be reported to interested constituencies. These decisions will, of necessity, be made at the school, district, or state level and will likely reflect educational, political, and economic considerations specific to the local context.

Expanded Table of Contents *

Long Tasks	Task Type	Circumstances of Performance
1. Anyone for Tennis?	45-minute planning task; nonroutine math results and student-life context; applied power	Students discuss task in pairs and then submit individual responses.
2. Making a Shed	45-minute design task; nonroutine adult-life context; applied power	Students discuss task in pairs and then submit individual responses.
3. Finding the Star Number	45-minute problem; nonroutine pure mathematical context	Students discuss task in pairs and then submit individual responses.
4. Bolts and Nuts	45-minute problem; nonroutine context from adult life; illustrative application	Students discuss task in pairs and then submit individual responses.
5. Field Trip	45-minute recommendation task; student-life context; applied power	Students discuss task in pairs and then submit individual responses.
6. Old Ruins	45-minute design task; illustrative application in an adult-life context	Students work individually.

* For explanations of terms that may be unfamiliar, see the Glossary and the *Dimensions of Balance* table in the Introduction.

Mathematical Content	Mathematical Processes
Other Mathematics with Data Handling: using logic and tables to combine constraints, then combinatorics to plan a schedule	formulation of a table from the given constraints; creation of a plan for the tournament; inference of a schedule from this information
Geometry, Shape, and Space with Number: build a 3-D model from a given net; measure model in centimeters and use 1:1 scale to describe full-size shed	manipulation of a model, measurements, and scaling; report of the description as an advertisement
Patterns, Functions, and Algebra: number exercise, finding and describing patterns as simple functions, with a table; generalize into rules, forward and reverse reasoning	manipulation of numbers; inference of the patterns and functions; formulation of functional relationships and predictions
Patterns, Functions, and Algebra with Number: measure a screw motion, number of turns in 25 mm; use proportion to predict for different lengths or number of turns	measurement and counting; formulation of functional relationships and predictions
Handling Data, Statistics, and Probability, with Number: combine class survey data given in three forms; graph and interpret it; compute the cost per student	transformation and manipulation of data; inference of preferred trip; formulation and manipulation of the cost model; report of the recommendations
Geometry, Shape, and Space: use line or mirror symmetry to complete designs; use properties of squares, triangles, and parallelograms to describe the designs	inference, transformation, and manipulation of the shapes from the symmetry; formulation to describe them and communicate it clearly

Expanded Table of Contents

Short Tasks	Task Type	Circumstances of Performance
7. Millie's Business	15-minute planning task; nonroutine context from adult life; applied power	Students work individually.
8. Seating Groups	15-minute problem; pure mathematics in a non-routine context from student life	Students work individually.
9. Fourth Graders	15-minute open-ended exercise; student-life context; applied power	Students work individually.
10. Rotating Shapes	5-minute technical exercise	Students work individually.
11. Going to Grandma's	5-minute technical exercise	Students work individually.
12. Drawing a Spinner	15-minute problem; nonroutine in context and mathematical demand	Students work individually.
13. Addition Rings	20-minute problem; nonroutine context and mathematical demand	Students work individually.
14. Anthony Pours Juice	5-minute technical exercise	Students work individually.
15. Tile Pattern	5-minute problem	Students work individually.
16. Danville's Library	5-minute problem; real-world context	Students work individually.

Mathematical Content	Mathematical Processes
Number and Quantity: use simple fractions, and rates to figure out which of two kinds of supplies will be used up first, given the amount on hand and the amount used each day	formulation of the approach; transformation and manipulation of the given quantities; inference
Other Mathematics: logic and combinatorics, listing all possible permutations of four children subject to given constraints	formulation of a systematic approach; manipulation of the permutations to carry it through
Handling Data, Statistics, and Probability: interpret data in a double bar graph about 3 classes; suggest comparisons from the data	formulation of interesting questions; inference of the answers from the data
Geometry, Space, and Shape: transformation and symmetry on a grid; congruency	manipulation—rotation of a given figure about a symmetry point
Geometry, Space, and Shape: location and movement on a grid	manipulation—maps, scales, and compass points
Data, Statistics, and Probability: basic concepts of statistics and probability to design the spinner from given tally count	interpretation of data; formulation of a strategy for use in a design; computation of the proportions; explanation
Number and Quantity: basic concepts of whole number; simple and reverse operations; representation	mainly manipulation; the nonroutine format requires some formulation of approach and inference
Number and Quantity: basic concepts of fractions; measurement or estimation	formulation in comparing before and after pictures; manipulation
Geometry, Space, and Shape: transformation of properties of shapes – translation (also a rotation solution)	balance of inference, formulation, and transformation through manipulation of the shapes involved
Data, Statistics, and Probability: representation of number data in a pictogram	interpretation of some data; inference on the representation; transformation of the other data

Task 1

Overview

Organize information.

Translate information from paragraph form into a table.

Create a schedule using data from the table.

Anyone for Tennis?

Long Task

Task Description

In this task, students are provided with information about the days of the week on which each of four people is able to play tennis. Students are asked to translate this written information into a table. They are then asked to plan a tournament using this information.

Assumed Mathematical Background

It is assumed students have some experience with organizing information and displaying information in tabular form.

Core Elements of Performance

- organize information
- translate information from paragraph form into a table
- create a schedule using data from the table

Circumstances

Grouping:	Students may discuss the task in pairs, but each student should complete an individual written response.
Materials:	rulers
Estimated time:	45 minutes

Name Date

Anyone for Tennis?

This problem gives you the chance to

- *organize information*
- *display information in a table*
- *use your table to plan a tournament*

Jane, Sam, Alan, and Molly like to play tennis together.

Jane is not able to play on Tuesday, Wednesday, and Saturday.

Sam can play on Monday, Wednesday, and Thursday.

Alan cannot play on Monday and Thursday.

Molly can play on Monday, Tuesday, and Friday.

None of them can play on Sunday.

1. Make a table showing which days of the week each student can play tennis.

© The Regents of the University of California

Name Date

2. Are there any days when all four students can play tennis? Explain your reasoning.

3. They decide to hold a tournament to see who is the best tennis player. On which days can more than one game be played?

4. Make a plan to show when each person can play every other person.

5. How many games will each person play?

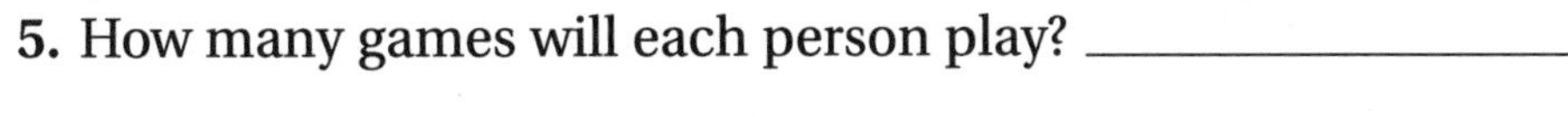

6. How many games will be played in the tournament? Explain how you figured it out.

© The Regents of the University of California

Task

A Sample Solution

1.

	Mon.	Tues.	Wed.	Thurs.	Fri.	Sat.	Sun.
Jane	X			X	X		
Sam	X		X	X			
Alan		X	X		X	X	
Molly	X	X			X		

The "X" indicates days when each student *can* play. Other types of tables are acceptable.

2. In the table, there are no days that show all four students with an "X." Therefore, there are no days on which all the students can play.

3. On Monday: Jane can play Sam; Sam can play Molly; Jane can play Molly. On Friday: Jane can play Alan; Jane can play Molly; Alan can play Molly.

4.

	Jane	Sam	Alan	Molly
Jane	X	Mon./Thurs.	Fri.	Mon./Fri.
Sam	Mon./Thurs.	X	Wed.	Mon.
Alan	Fri.	Wed.	X	Tues./Fri.
Molly	Mon./Fri.	Mon.	Tues./Fri.	X

Tables or lists based on days of the week are also acceptable.

5. Each person will play 3 games (for example, Jane must play Sam, Alan, and Molly).

6. There will be 6 games played in the tournament. (The order does not matter as long as each student plays every other student once.)
Jane must play Sam, Alan, and Molly—3 games.
Then Sam must play Alan and Molly—2 games.
Then Alan must play Molly—1 game.

Using this Task

Task

Read through the prompt with students to ensure that they understand the task and answer any questions that arise.

It may be necessary to explain to students that in a singles game there are only two players: each player competes with one other player.

It may also be necessary to explain that in a tournament each player plays every other player exactly once.

Extensions

You may wish to extend this task by using some of the ideas that follow.

- Several teams play in a competition. How many games are necessary for different numbers of teams? Can you generalize?
- Several teams want to arrange a baseball league subject to constraints on the days/dates when they are able to play. Figure out which days/dates are possible.
- Several teams play in a competition; provide a list of matches played and their results. If a win gets 2 points, a draw gets 1 point, and a loss gets 0 points, who is top of the league? But this is not the only way to allocate points. Can students suggest a better/different way? Does using this new way of allocating points change the team at the top of the league?

Task

Characterizing Performance

This section offers a characterization of student responses and provides indications of the ways in which the students were successful or unsuccessful in engaging with and completing the task. The descriptions are keyed to the *Core Elements of Performance.* Our global descriptions of student work range from "The student needs significant instruction" to "The student's work meets the essential demands of the task." Samples of student work that exemplify these descriptions of performance are included below, accompanied by commentary on central aspects of each student's response. These sample responses are *representative;* they may not mirror the global description of performance in all respects, being weaker in some and stronger in others.

The characterization of student responses for this task is based on these *Core Elements of Performance:*

1. Organize information.
2. Translate information from paragraph form into a table.
3. Create a schedule using data from the table.

Descriptions of Student Work

The student needs significant instruction.

Responses at this level have limited success in one or two of the core elements of performance. Typically they make an attempt to organize the data from the paragraph into some form of table. The table, however, may contain errors. Or the response uses the information from the paragraph to decide whether there are any days when all four can play tennis together.

Student A

This response has organized the information from the paragraph into a table that contains minor errors. There is no "X" in the box corresponding to Jane/Saturday. No other part of the task has been attempted.

The student needs some instruction.

Responses at this level organize the information provided to make a table showing the days on which each student can play tennis. They typically use

the table to decide whether there are days when all four can play tennis together. Other aspects of the tasks are either not attempted or are incorrect.

Student B

This response contains a correct table. The table is used to respond correctly to question 2. The explanation, however, is not specific and the rest of the task is not attempted.

The student's work needs to be revised.

Responses at this level demonstrate some success with all the core elements of performance. The table is complete and correct or nearly so and it is used to answer question 2 correctly. These responses also contain a workable schedule that contains only minor errors.

Student C

This response shows a well-organized table. The table is used to respond to question 2 and the student explains what the symbols represent. The table is used to respond to question 3, but this student has misinterpreted question 4, apparently reading it as a follow-up to question 3. The responses to questions 5 and 6 are incorrect.

The student's work meets the essential demands of the task.

Responses at this level satisfactorily complete almost all aspects of the core elements of performance. They organize the information provided to make a table showing on which days each student can play tennis. They use the table to decide whether there are days when all four can play tennis together. They make a schedule for a tournament, showing on which day each person can play every other person and explaining how many games will be played in the tournament.

Student D

This response shows a clearly organized table. The table is used to answer question 2. It is also used to make a schedule showing on which days each person can play every other person (incomplete for Thursday and Friday). Finally, the response correctly answers questions 5 and 6.

Student A

Anyone for Tennis?

This problem gives you the chance to

- *organize information*
- *display information in a table*
- *use your table to plan a tournament*

Jane, Sam, Alan, and Molly like to play tennis together.
Jane is not able to play on Tuesday, Wednesday, and Saturday.
Sam can play on Monday, Wednesday, and Thursday.
Alan cannot play on Monday and Thursday.
Molly can play on Monday, Tuesday, and Friday.
None of them can play on Sunday.

1. Make a table showing which days of the week each student can play tennis.

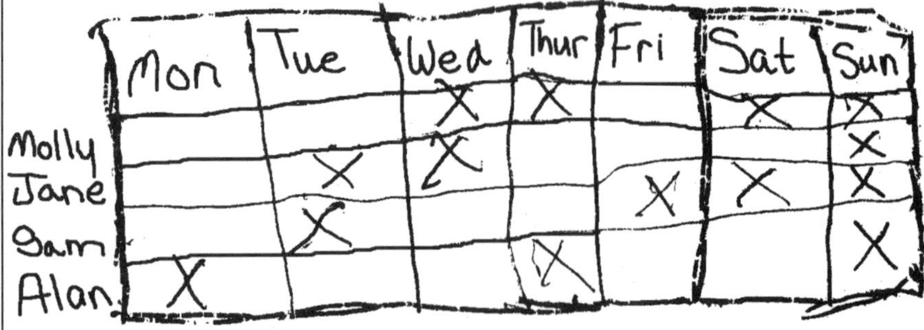

	Mon	Tue	Wed	Thur	Fri	Sat	Sun
			X	X		X	X
Molly		X	X				X
Jane					X	X	X
Sam		X					X
Alan	X			X			

Student B

Anyone for Tennis?

This problem gives you the chance to

- *organize information*
- *display information in a table*
- *use your table to plan a tournament*

Jane, Sam, Alan, and Molly like to play tennis together.

Jane is not able to play on Tuesday, Wednesday, and Saturday.

Sam can play on Monday, Wednesday, and Thursday.

Alan cannot play on Monday and Thursday.

Molly can play on Monday, Tuesday, and Friday.

None of them can play on Sunday.

1. Make a table showing which days of the week each student can play tennis.

names	Mon	tues	weds	thurs	fri	Sat	Sun
Jane	Y	N	N	Y	Y	N	N
Sam	Y	N	Y	Y	N	N	N
Alan	N	Y	Y	N	Y	Y	N
molly	Y	Y	N	N	Y	N	N

2. Are there any days when all four students can play tennis? Explain your reasoning. No, look at the chart

Student C

Anyone for Tennis?

This problem gives you the chance to

- *organize information*
- *display information in a table*
- *use your table to plan a tournament*

Jane, Sam, Alan, and Molly like to play tennis together.

Jane is not able to play on Tuesday, Wednesday, and Saturday.

Sam can play on Monday, Wednesday, and Thursday.

Alan cannot play on Monday and Thursday.

Molly can play on Monday, Tuesday, and Friday.

None of them can play on Sunday.

1. Make a table showing which days of the week each student can play tennis.

	SU	MON	TUES	WED	THU	FRI	SAT.
Jane	X	•	X	X	•	•	X
Sam	X	•	X	•	•	X	X
Alan	X	X	•	•	X	•	•
molly	X	•	•	X	X	•	X

Student C

2. Are there any days when all four students can play tennis? Explain your reasoning. The x represents no, The • represents yes. All of them cannot play on the same day.

3. They decide to hold a tournament to see who is the best tennis player. On which days can more than one game be played? Friday, and Monday.

4. Make a plan to show when each person can play every other person.

mon-
Jane vs Sam
Jane vs molly
Sam vs molly

fri-
Jane vs Alan
Jane vs. Molly
Alan vs. molly

5. How many games will each person play? 2

6. How many games will be played in the tournament? 3
Explain how you figured it out. Each person would play 2 games a day. And there would be 3 games a day.

Student D

Anyone for Tennis?

This problem gives you the chance to

- *organize information*
- *display information in a table*
- *use your table to plan a tournament*

Jane, Sam, Alan, and Molly like to play tennis together.

Jane is not able to play on Tuesday, Wednesday, and Saturday.

Sam can play on Monday, Wednesday, and Thursday.

Alan cannot play on Monday and Thursday.

Molly can play on Monday, Tuesday, and Friday.

None of them can play on Sunday.

1. Make a table showing which days of the week each student can play tennis.

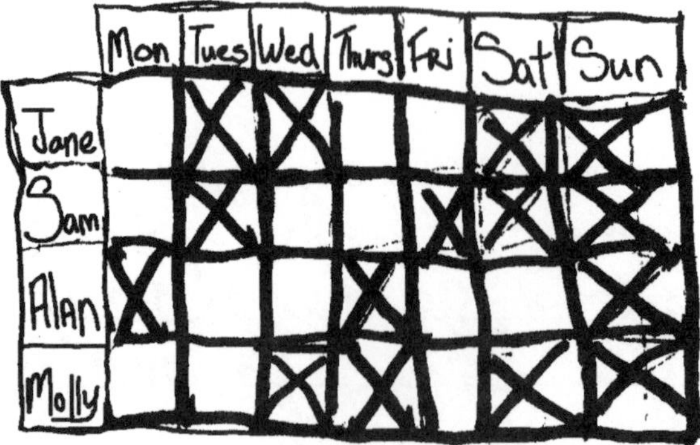

	Mon	Tues	Wed	Thurs	Fri	Sat	Sun
Jane		X	X			X	X
Sam		X			X	X	X
Alan	X			X			X
Molly			X	X		X	X

Student D

2. Are there any days when all four students can play tennis? Explain your reasoning. No, one or more people are busy every day.

3. They decide to hold a tournament to see who is the best tennis player. On which days can more than one game be played? Monday and Fri.

4. Make a plan to show when each person can play every other person.

Mon.
Jane + Sam
Jane + Molly
Sam + Molly

Tues.
Alan + Molly

Wed.
Sam + Alan

Thurs.

Fri.
Jane + Alan

5. How many games will each person play? 3

6. How many games will be played in the tournament? 6
Explain how you figured it out. On: Mon - 3 games Tues - 1 Wed. - 1 Thurs - 0 Fri - 1 Sat - 0 Sun - 0 We transfered info from first chart.

3
+ 1
+ 1
+ 1
6

Task 2

Making a Shed

Overview

Make a 3-D model from a net.

Measure in centimeters.

Use scale to figure actual size of shed.

Use mathematical language to describe the real shed.

Long Task

Task Description

Students are provided with the net of a shed that they make into a three-dimensional model. They measure the model and convert the measurements to find the actual size of the shed. Then they write an advertisement for a shed like the model.

Assumed Mathematical Background

The task involves spatial reasoning, measurement, and a little number work. Students are expected to have some prior experience with measuring in centimeters and using scale drawings.

Core Elements of Performance

- make a three-dimensional model from a net
- measure the model in centimeters
- use scale to figure out the size of the real shed
- use mathematical language to describe the real shed

Circumstances

Grouping:	Students may discuss the tasks in pairs, but each student should complete an individual written response.
Materials:	scissors, tape, and a centimeter ruler
Estimated time:	45 minutes

Name Date

Making a Shed

This problem gives you the chance to

- *make a model shed*
- *measure your model accurately*
- *figure out the measurements of the real shed*
- *write an advertisement for a real shed*

1. Make a model shed using the Resource Sheet on the next page.
2. Write an advertisement to help sell sheds like this. In your ad you need to specify the size of the shed, the size of the door, the number of windows, the size of the windows, and the shapes of the windows. (*Remember, 1 centimeter on the model represents 1 foot on the real shed.*) ____________________

© The Regents of the University of California

Name Date

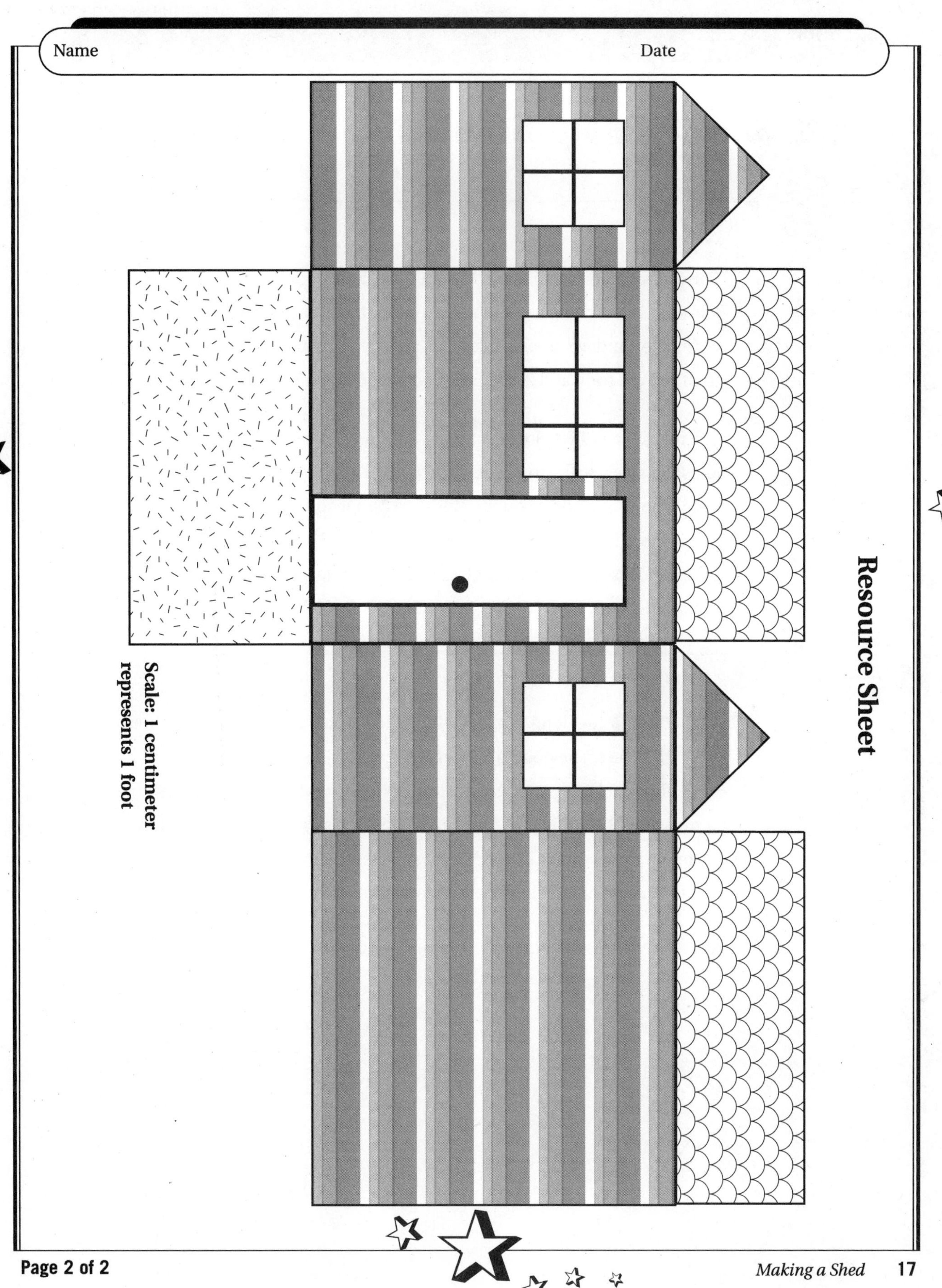

© The Regents of the University of California

Task

A Sample Solution

The model shed is 7 cm long, $3\frac{1}{2}$ cm wide, and 7 cm high, where the roof joins the sides, rising to 9 cm at the peak. The door is 6 cm high and 2 cm wide. The windows on the sides are 2 cm by 2 cm and the window on the front is 3 cm wide and 2 cm high. Each window pane is 1 square centimeter.

The real shed is 7 feet long, $3\frac{1}{2}$ feet wide, and 7 feet high where the roof joins the sides, rising to 9 feet at the peak. The door is 6 feet high and 2 feet wide. The two windows on the sides are 2-foot squares and the rectangular window on the front is 3 feet wide and 2 feet high. The windows on the sides each need 4 square pieces of glass and the window on the front needs 6 square pieces of glass: all the pieces of glass are 1-foot squares.

Sheds for sale

Sheds are 7 feet long, $3\frac{1}{2}$ feet wide, minimum height 7 feet, maximum height 9 feet. The door is 6 feet high, 2 feet wide. On the front of the shed is one large rectangular window (made with 6 panes of glass) that measures 3 feet long by 2 feet high. On each side is a square window (made with 4 panes of glass). The square windows measure 2 feet by 2 feet. All panes of glass are 1 square foot.

Using this Task

Task 2

Read through the prompt with your students to ensure that they understand the task, and answer any questions that arise. Be sure that students understand what a net is and how to use it to make their model shed.

Explain that making the model shed is only the first stage of the task. They also need to figure out the measurements of the real shed and then write an advertisement that tells readers the exact size of the shed, the door, and the windows.

Extensions

You may wish to extend this task by using some of the ideas that follow.

- Fred wants a shed in which he can keep his garden tools and a lawn mower.
  Is the shed in this task a suitable size?
  If not, decide what size shed he needs and draw a sketch.
  Decide on a scale and draw the net of this shed.
  Use the net to make a model.
- Your family wants a shed in which to keep bicycles and other sports gear.
  Is the shed in this task a suitable size?
  If not, decide what size shed is needed and draw a sketch.
  Decide on a scale and draw the net of this shed.
  Use the net to make a model.
- Select a car and find its dimensions.
  Decide what size garage is suitable for the car and draw a sketch.
  Decide on a scale and draw the net of a suitable garage.
  Use the net to make a model.
- You want to paint your shed.
  Figure out the surface area of all four walls of the shed.
  Then figure out the floor area of the shed.
  How big a carpet is needed to cover the entire floor?

Task

Characterizing Performance

This section offers a characterization of student responses and provides indications of the ways in which the students were successful or unsuccessful in engaging with and completing the task. The descriptions are keyed to the *Core Elements of Performance.* Our global descriptions of student work range from "The student needs significant instruction" to "The student's work meets the essential demands of the task." Samples of student work that exemplify these descriptions of performance are included below, accompanied by commentary on central aspects of each student's response. These sample responses are *representative;* they may not mirror the global description of performance in all respects, being weaker in some and stronger in others.

The characterization of student responses for this task is based on these *Core Elements of Performance:*

1. Make a three-dimensional model from a net.
2. Measure the model in centimeters.
3. Use scale to figure out the size of the real shed.
4. Use mathematical language to describe the real shed.

Descriptions of Student Work

The student needs significant instruction.

Responses at this level show some success in at least one core element of performance. Typically they cut out the net and make a model shed. The response may show an attempt to write an advertisement that may not contain any information about the size of the shed, windows, or door.

Student A

This student made a model shed, but there is no evidence that any part of it was measured. The advertisement does not contain mathematical language or any information about size.

Task

The student needs some instruction.

Responses at this level show some success in at least two of the core elements of performance. The shed is accurately constructed, and there is evidence that some measurements were made. The advertisement contains some information relating to size and shape, but this may not be accurate or complete.

Student B

This response shows evidence that some measurements of the shed were made, and the advertisement contains some information about size, but all of this is inaccurate.

The student's work needs to be revised.

Responses at this level show some evidence of success in at least three of the core elements of performance. The advertisement contains most of the dimensions of the shed, windows, and door. These responses may omit one or two measurements, or one or two measurements may be inaccurate. There is, however, clear evidence that some measurements were made accurately and that the scale was used correctly in figuring out the actual size of the shed.

Student C

This response contains some accurate information about the length and width of the real shed. Measurements of doors and windows as well as the height of the shed are missing. The response does provide evidence that some measurements were accurately made and that the scale was accurately used to figure actual size.

The student's work meets the essential demands of the task.

Responses at this level show evidence of substantial success in all four core elements of performance. There is evidence that most measurements were made accurately and that the scale was used to convert the measurements of the model to measurements of the shed.

Student D

The advertisement in this response contains accurate information about the length and width of the shed and the size of the door and the windows. The student has left out the number of windows and the height of the shed, but shows a clear understanding of the skills needed to accomplish the task.

Student A

Making a Shed

This problem gives you the chance to

- *make a model shed*
- *measure your model accurately*
- *figure out the measurements of the real shed*
- *write an advertisement for a real shed*

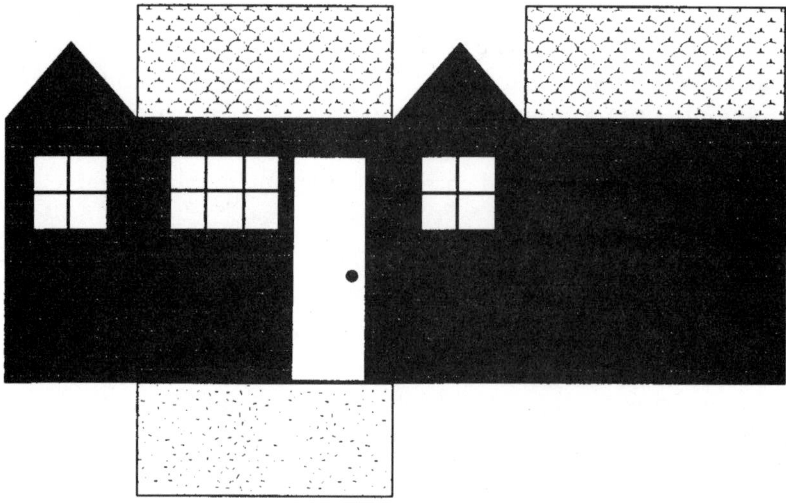

1. Make a model shed using the Resource Sheet on the next page.
2. Write an advertisement to help sell sheds like this. In your ad you need to specify the size of the shed, the size of the door, the number of windows, the size of the windows, and the shapes of the windows. (*Remember, 1 centimeter on the model represents 1 foot on the real shed.*) Hurry, hurry, Hurry Do you want a shed? On the outside of the shed the color is yellow and the inside is white, but if you don't like these colors you can change them.

Student B

Making a Shed

This problem gives you the chance to

- *make a model shed*
- *measure your model accurately*
- *figure out the measurements of the real shed*
- *write an advertisement for a real shed*

1. Make a model shed using the Resource Sheet on the next page.
2. Write an advertisement to help sell sheds like this. In your ad you need to specify the size of the shed, the size of the door, the number of windows, the size of the windows, and the shapes of the windows. (*Remember, 1 centimeter on the model represents 1 foot on the real shed.*) SHEDS FOR SALE
3 Feet wide and 7½ feet long.
The price is 1,000 dollars

Student C

Making a Shed

This problem gives you the chance to

- *make a model shed*
- *measure your model accurately*
- *figure out the measurements of the real shed*
- *write an advertisement for a real shed*

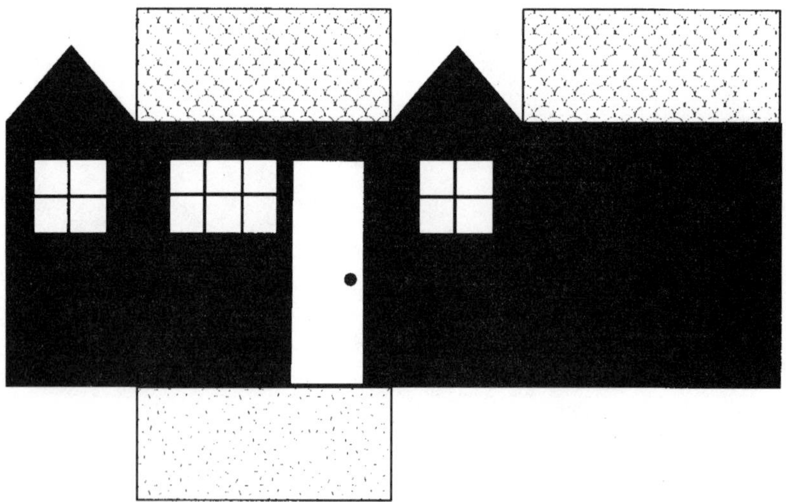

1. Make a model shed using the Resource Sheet on the next page.
2. Write an advertisement to help sell sheds like this. In your ad you need to specify the size of the shed, the size of the door, the number of windows, the size of the windows, and the shapes of the windows. (*Remember, 1 centimeter on the model represents 1 foot on the real shed.*) Sheds for sale They are 7 feet long and 3 1/2 feet wide. They have a big window at the front and small side window.

Student D

Making a Shed

This problem gives you the chance to

- *make a model shed*
- *measure your model accurately*
- *figure out the measurements of the real shed*
- *write an advertisement for a real shed*

1. Make a model shed using the Resource Sheet on the next page.
2. Write an advertisement to help sell sheds like this. In your ad you need to specify the size of the shed, the size of the door, the number of windows, the size of the windows, and the shapes of the windows. (*Remember, 1 centimeter on the model represents 1 foot on the real shed.*) Shed For Sale

Size - 3½ Feet x 7 Feet long
Color - Brown
Windows - 3 Feet long x 2 Feet wide
2 Feet x 2 Feet
Door - 6 feet higl x 2 Feet wide

Task 3

Finding the Star Number

Overview

Use simple functions involving addition.

Complete a table of results.

Find a functional relationship.

Make predictions.

Long Task

Task Description

Students are given beginning numbers and rules that enable them to find the star number. They are asked to make a T-table showing the sums of beginning numbers and star numbers. Using this table, they are asked to predict the star number for a given sum of beginning numbers. Then they are asked to figure out what the beginning numbers might be for a given star number. Finally, they are asked to write down the relationship between the sum of the beginning numbers and the star number.

Assumed Mathematical Background

Students should have prior experience with simple functions, adding numbers using a calculator, finding relationships, and making predictions.

Core Elements of Performance

- use simple functions involving addition
- complete a table of results
- state the rule for a functional relationship
- make predictions

Circumstances

Grouping:	Students may discuss the task in pairs, but each student should complete an individual written response.
Materials:	calculators (optional)
Estimated time:	45 minutes

Name Date

Finding the Star Number

This problem gives you the chance to

- *look for number patterns*
- *use the patterns to make predictions about the star number*
- *describe the relationship between the star number and the beginning numbers*

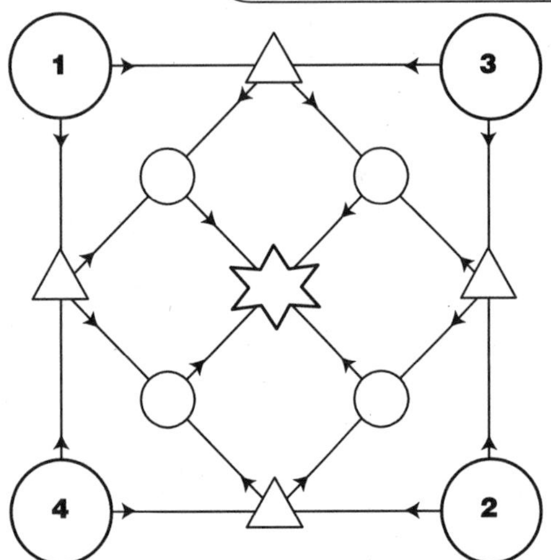

Rules for playing *Finding the Star Number:*

- Add the pairs of numbers in the circles on the same sides of the large square. (For example; 1 + 3 = 4, 1 + 4 = 5, 3 + 2 = 5, 4 + 2 = 6.)
- Write their sums in the triangles between them.
- Add the numbers in the triangles on the same sides of the small square.
- Write their sums in the circles between them.
- Add the four numbers in the small circles.
- Write this number inside the star.

1. Use the above rules to find the star number.

   a. What is the sum of the four beginning numbers (1 + 3 + 4 + 2)? ________

   b. What is the star number? ________

© The Regents of the University of California

Name Date

Now find each of these star numbers:

2.

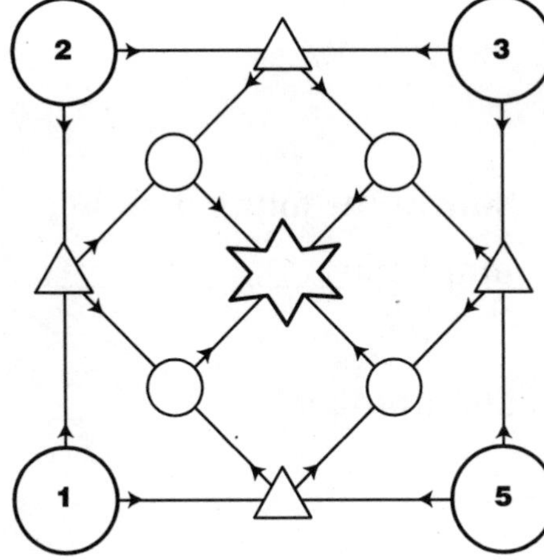

a. **Sum of the four beginning numbers** ____________

b. **Star number** __________

3.

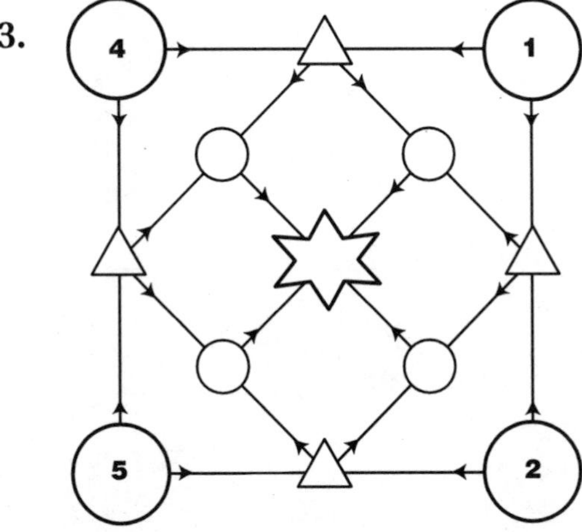

a. **Sum of the four beginning numbers** ____________

b. **Star number** __________

4.

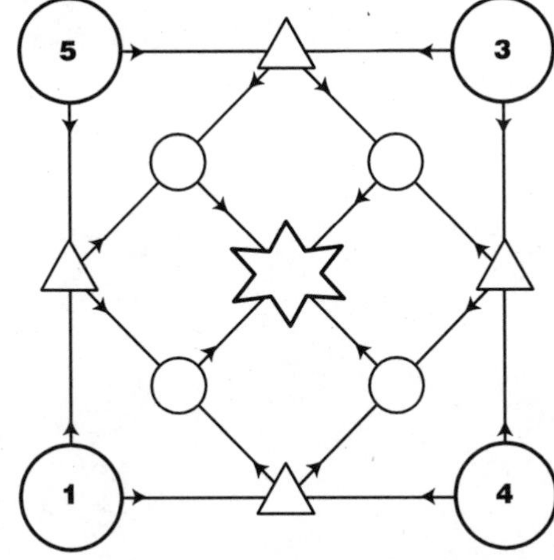

a. **Sum of the four beginning numbers** ____________

b. **Star number** __________

© The Regents of the University of California

Name Date

Now make up your own sets of four beginning numbers and figure out the star numbers for them.

5.

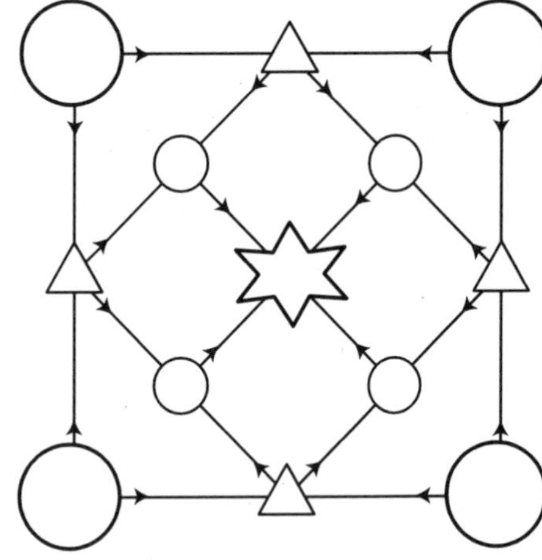

a. **Sum of the four beginning numbers** ___________

b. **Star number** ___________

6.

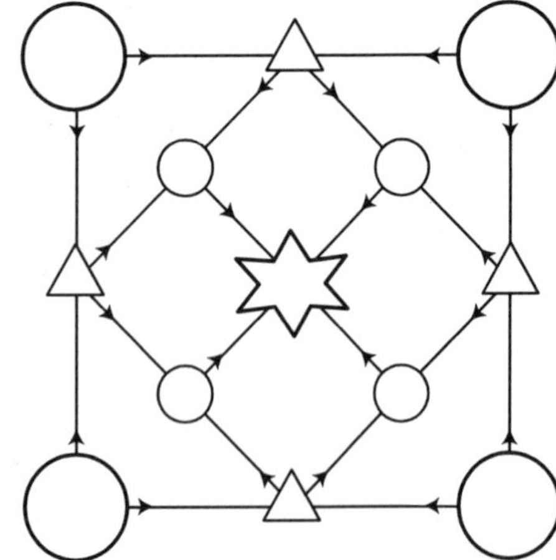

a. **Sum of the four beginning numbers** ___________

b. **Star number** ___________

© The Regents of the University of California

Name Date

7. Fill in this table by recording the sum of the four beginning numbers and the star number for each star number square. The first one is done for you.

Sum of beginning numbers	Star number
10	40

8. What rule would help you predict the star number if you know the sum of the four beginning numbers? ____________

____

____

© The Regents of the University of California

Name Date

9. Use the table to predict the star number when the sum of the beginning numbers is 20. ______

10. What do you think the beginning numbers can be when the star number is 48? ______

11. What other sets of numbers could be the beginning numbers for a star number of 48? How do you know? ______

© The Regents of the University of California

A Sample Solution

Task 3

1.

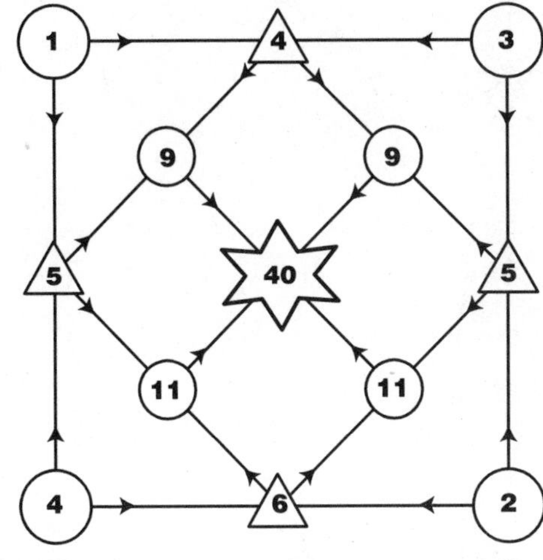

a. Sum of the four beginning numbers 10

b. Star number 40

2.

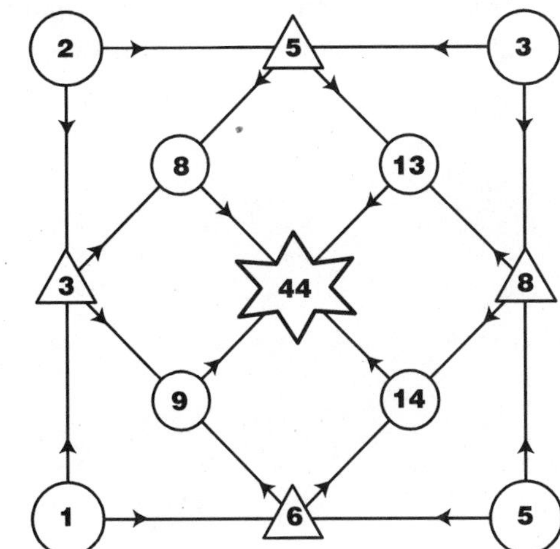

a. Sum of the four beginning numbers 11

b. Star number 44

3.

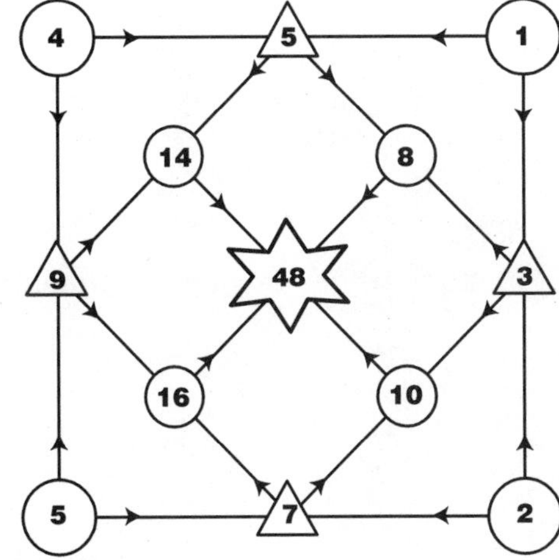

a. Sum of the four beginning numbers 12

b. Star number 48

Task 3

4.

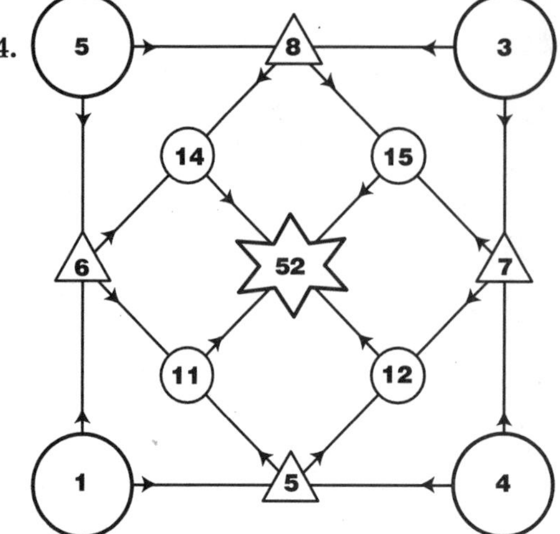

a. Sum of the four beginning numbers 13

b. Star number 52

The last two star number problems (5 and 6) will vary from response to response. Because of this variation, the last two entries in the table will also vary from response to response. In this table we show the star number for all sums of beginning numbers from 10 to 20.

7.

Sum of beginning numbers	Star number
10	40
11	44
12	48
13	52
14	56
15	60
16	64
17	68
18	72
19	76
20	80

8. You can predict the star number if you multiply the sum of the beginning numbers by four.

Task

9. When the sum of the beginning numbers is 20, the star number is 80.

10. From the table, if the star number is 48, the sum of the beginning numbers is 12.

11. The beginning numbers can be 1, 2, 4, 5; or 1, 2, 3, 6; or any other set of four numbers whose sum is 12.

Task

Using this Task

Read through the prompt with your students to ensure that they understand the task and answer any questions that arise. Work through the puzzle on page 28 with the whole class on an overhead or on the board. Explain that finding the star number is only the first step in solving this puzzle. After finding a few star numbers, students should look for patterns in the relationship of the four beginning numbers to the star number. They should try to find a rule for finding the star number for a given sum of beginning numbers. They should also find a rule for the sum of beginning numbers for a given star number.

Extensions

Ask students if they can figure out why the star number is four times the sum of the beginning number. You may find it helpful to provide several piles of objects representing each beginning number as students try to figure out the connection.

Students can explore similar puzzles. Below are two possibilities; one of them is much simpler than the star number puzzle and one more difficult.

In this example, the "star" number is twice the sum of the beginning numbers.

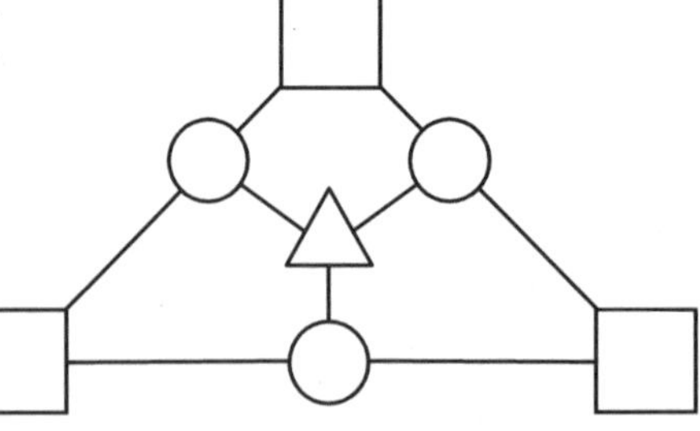

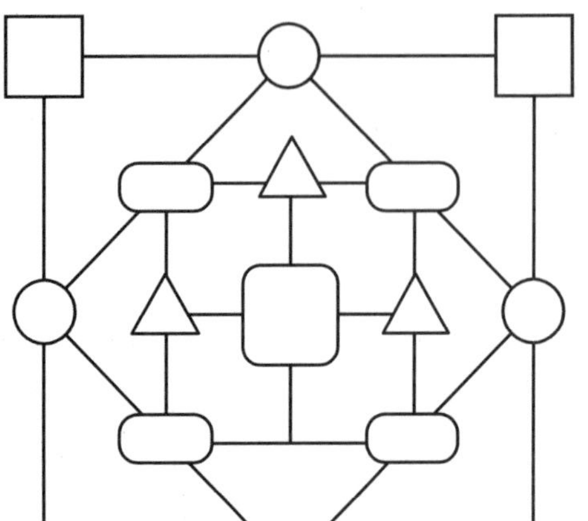

In this example, the "star" number is eight times the sum of the beginning numbers.

Characterizing Performance

Task

This section offers a characterization of student responses and provides indications of the ways in which the students were successful or unsuccessful in engaging with and completing the task. The descriptions are keyed to the *Core Elements of Performance.* Our global descriptions of student work range from "The student needs significant instruction" to "The student's work meets the essential demands of the task." Samples of student work that exemplify these descriptions of performance are included below, accompanied by commentary on central aspects of each student's response. These sample responses are *representative;* they may not mirror the global description of performance in all respects, being weaker in some and stronger in others.

The characterization of student responses for this task is based on these *Core Elements of Performance:*

1. Use simple functions involving addition.
2. Complete a table of results.
3. State the rule for a functional relationship.
4. Make predictions.

Descriptions of Student Work

The student needs significant instruction.

These responses show evidence of an attempt to find the star numbers and to complete the table. There may be some errors in the star number puzzles, which would lead to incorrect entries in the table. Other parts of the task are either not attempted or attempted unsuccessfully.

Student A

This response attempts to find the star numbers, but addition mistakes are made in two of the puzzles. The T-table is filled in, including the addition errors from the puzzle. No attempt is made to state a rule governing the functional relationship. The response does give the correct star number for a sum of beginning numbers of 20. In this case, one of the problems that the student created in question 6 had beginning numbers that added up to 80.

Task

The student needs some instruction.

These responses find the star numbers from the numbers given and from their own numbers. The T-table is completed correctly, but other aspects of the task are either not attempted or are incorrect.

Student B

This response correctly finds all the star numbers and fills the table in correctly with only one incorrect entry, the result of miscopying. The response makes an attempt to find patterns in the T-table, but is not able to articulate the rule. No other part of the task is attempted.

The student's work needs to be revised.

These responses find the star numbers from the numbers given and from their own numbers. The T-table is completed correctly. The response shows some evidence of understanding the functional relationship between the sum of the beginning numbers and the star number. This evidence may be in the form of stating the rule, or answering at least two of the three questions on the last page of the task.

Student C

This response correctly finds all the star numbers and fills in the table correctly. It states that the rule is counting by fours and gives the star number for a sum of beginning numbers of 20. (This response did not already have a sum of beginning number of 20 as one of the choices in problems 5 and 6.) The last two parts of the task are not attempted.

The student's work meets the essential demands of the task.

These responses find the star number from the numbers given and from their own numbers. The T-table is completed correctly. The rule governing the functional relationship between the star number and the four beginning numbers is clearly stated. Questions 9, 10, and 11 are answered. The response may or may not give a second set of beginning numbers for the star number of 48.

Student D

This response completes all aspects of the task correctly, except that it does not enter the results from the last two star number puzzles from problems 5 and 6 in the T-table. The rule is given as "× 4." Four beginning numbers for a star number are given, and an alternative set of beginning numbers is given.

Student A

Now find each of these star numbers:

2.

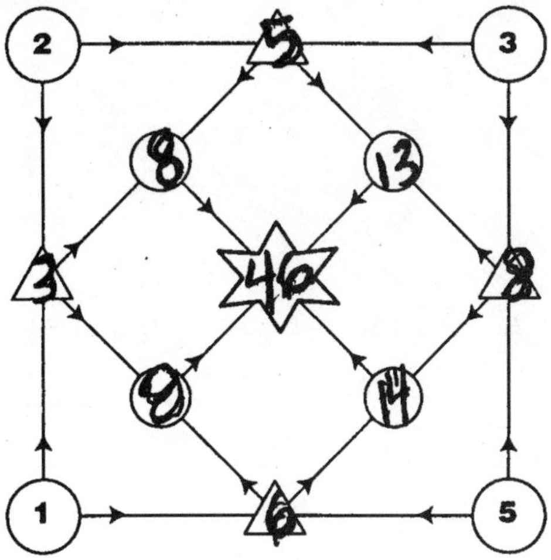

a. Sum of the four beginning numbers 11

b. Star number 46

3.

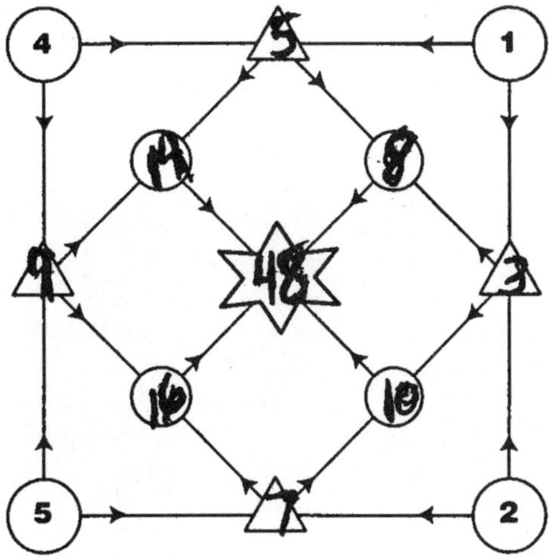

a. Sum of the four beginning numbers 12

b. Star number 48

4.

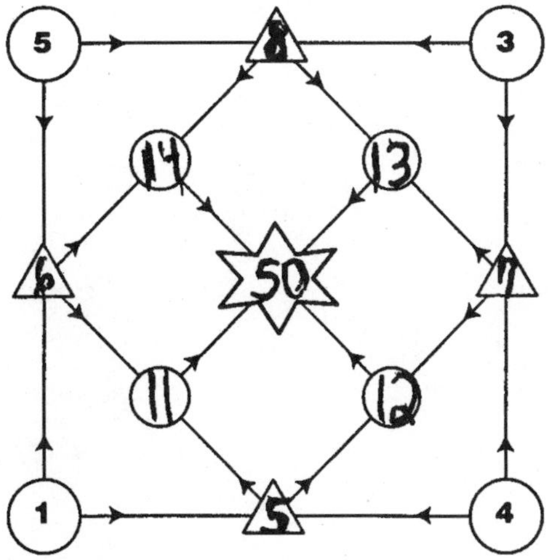

a. Sum of the four beginning numbers 13

b. Star number 50

Student A

Now make up your own sets of four beginning numbers and figure out the star numbers for them.

5.

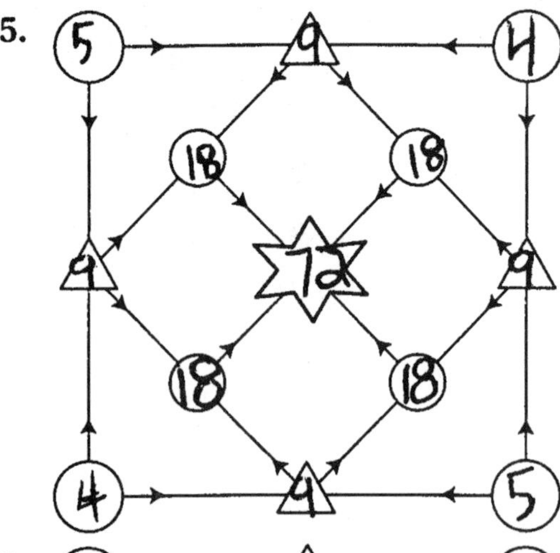

a. Sum of the four beginning numbers 18

b. Star number 72

6.

a. Sum of the four beginning numbers 20

b. Star number 80

Student A

7. Fill in this table by recording the sum of the four beginning numbers and the star number for each star number square. The first one is done for you.

Sum of beginning numbers	Star number
10	40
11	46
12	48
13	50
18	72
20	80

8. What rule would help you predict the star number if you know the sum of the four beginning numbers? ______________

____

____

Student A

9. Use the table to predict the star number when the sum of the beginning numbers is 20.

if you started with 20 you would get 80.

10. What do you think the beginning numbers can be when the star number is 48?

11. What other sets of numbers could be the beginning numbers for a star number of 48? How do you know?

Finding the Star Number ■ Student Work

Student B

Now find each of these star numbers:

2.

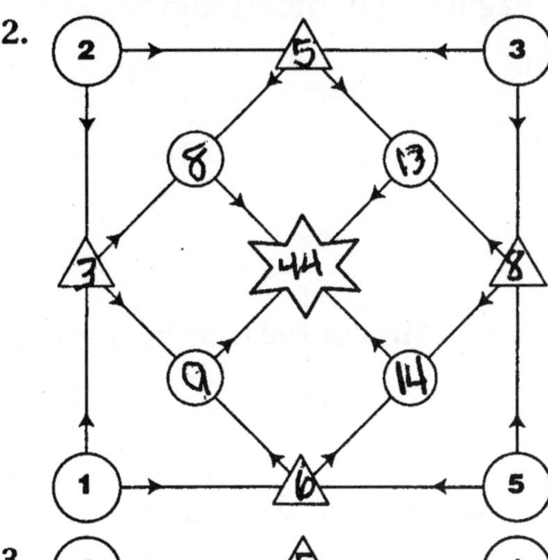

a. Sum of the four beginning numbers 11

b. Star number 44

3.

a. Sum of the four beginning numbers 12

b. Star number 48

4.

a. Sum of the four beginning numbers 13

b. Star number 52

Student B

Now make up your own sets of four beginning numbers and figure out the star numbers for them.

5.

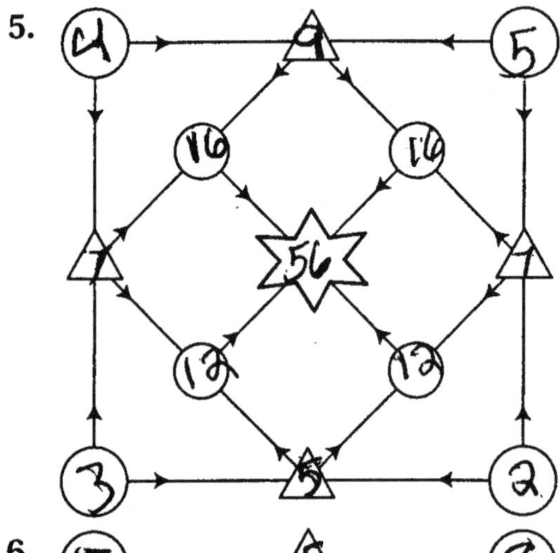

a. Sum of the four beginning numbers 14

b. Star number 56

6.

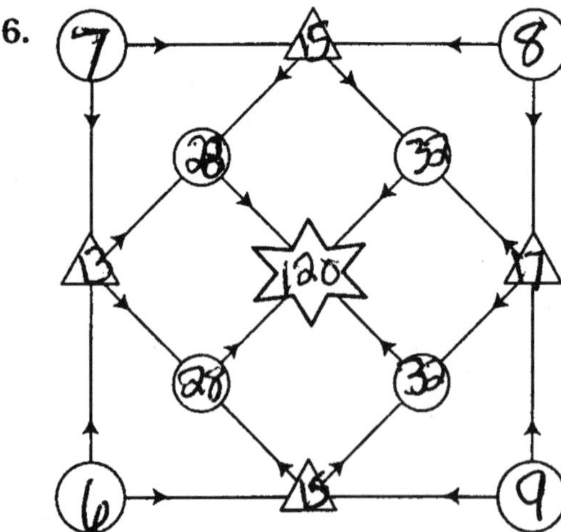

a. Sum of the four beginning numbers 30

b. Star number 120

Student B

7. Fill in this table by recording the sum of the four beginning numbers and the star number for each star number square. The first one is done for you.

Sum of beginning numbers	Star number
10	40
11	44
12	48
13	52
14	62
30	120

8. What rule would help you predict the star number if you know the sum of the four beginning numbers? ____________

____

____

Student B

9. Use the table to predict the star number when the sum of the beginning numbers is 20. If you use the T table to Predict the star number when the sum of the beginning numbers is 20 the Star number will probley be 54

10. What do you think the beginning numbers can be when the star number is 48?

I Dont get it.

11. What other sets of numbers could be the beginning numbers for a star number of 48? How do you know?

Student C

Now find each of these star numbers:

2.

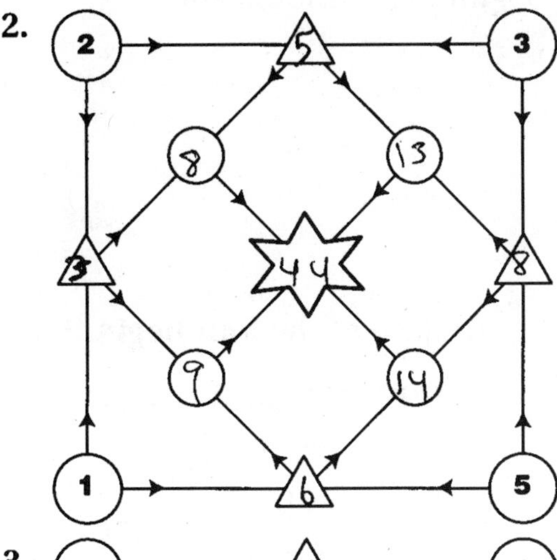

a. Sum of the four beginning numbers 11

b. Star number 44

3.

a. Sum of the four beginning numbers 12

b. Star number 48

4.

a. Sum of the four beginning numbers 13

b. Star number 52

Student C

Now make up your own sets of four beginning numbers and figure out the star numbers for them.

5.

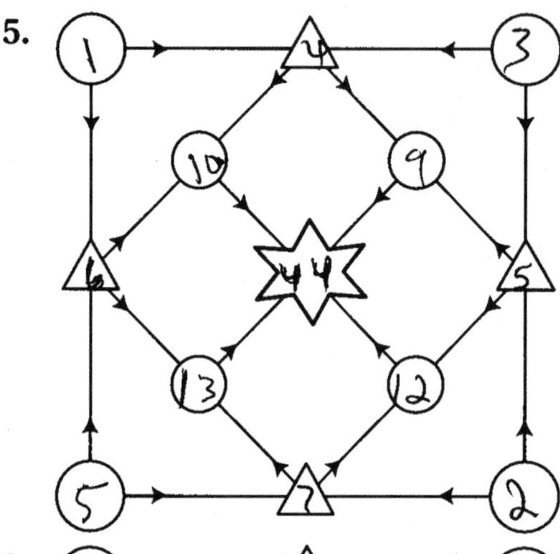

a. **Sum of the four beginning numbers** 11

b. **Star number** 44

6.

a. **Sum of the four beginning numbers** 12

b. **Star number** 48

Student C

7. Fill in this table by recording the sum of the four beginning numbers and the star number for each star number square. The first one is done for you.

Sum of beginning numbers	Star number
10	40
11	44
12	48
13	52
11	44
12	48

8. What rule would help you predict the star number if you know the sum of the four beginning numbers? I think the rule is that we are counting by fore's.

Student C

9. Use the table to predict the star number when the sum of the beginning numbers is 20. If I stared with 20, The star # would be 80.

10. What do you think the beginning numbers can be when the star number is 48?

I don't get it

11. What other sets of numbers could be the beginning numbers for a star number of 48? How do you know?

I dont get it.

Student D

Now find each of these star numbers:

2.

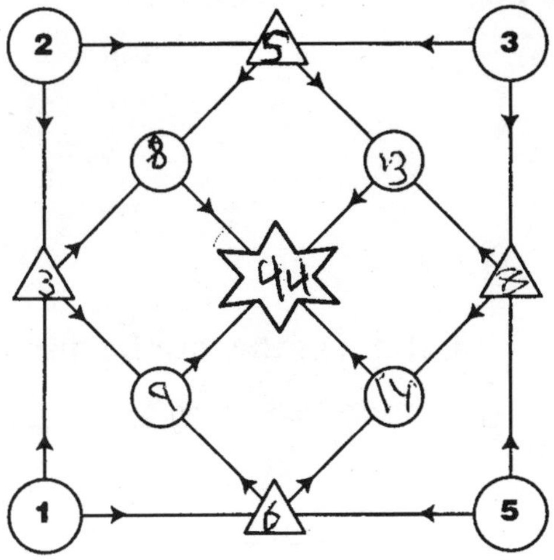

a. Sum of the four beginning numbers 11

b. Star number 44

3.

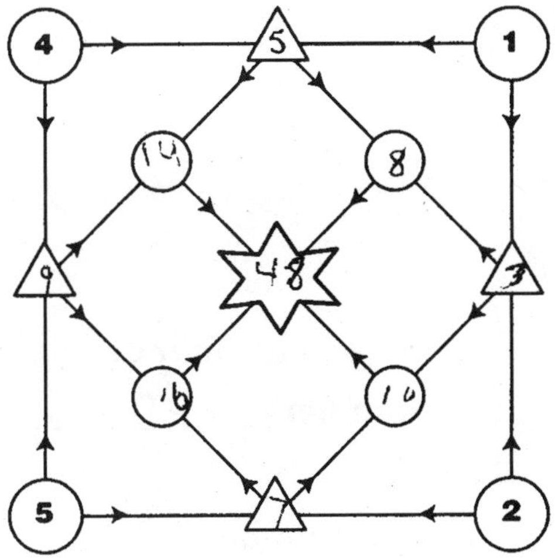

a. Sum of the four beginning numbers 12

b. Star number 48

4.

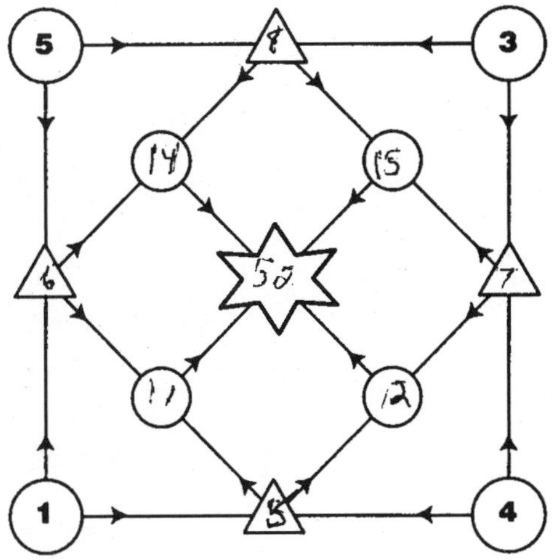

a. Sum of the four beginning numbers 13

b. Star number 52

Student D

Now make up your own sets of four beginning numbers and figure out the star numbers for them.

5.

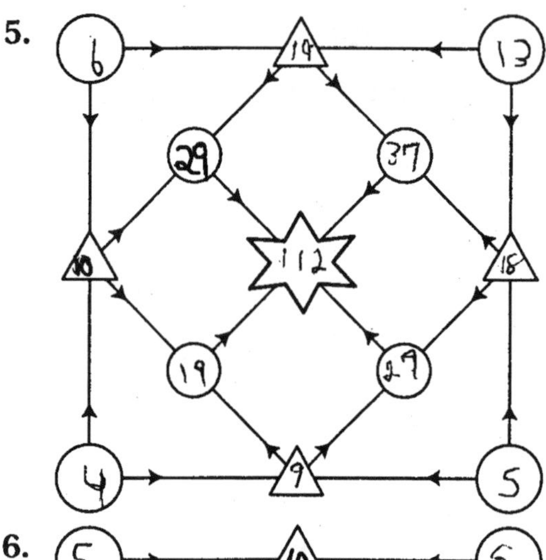

a. **Sum of the four beginning numbers** 28

b. **Star number** 112

6.

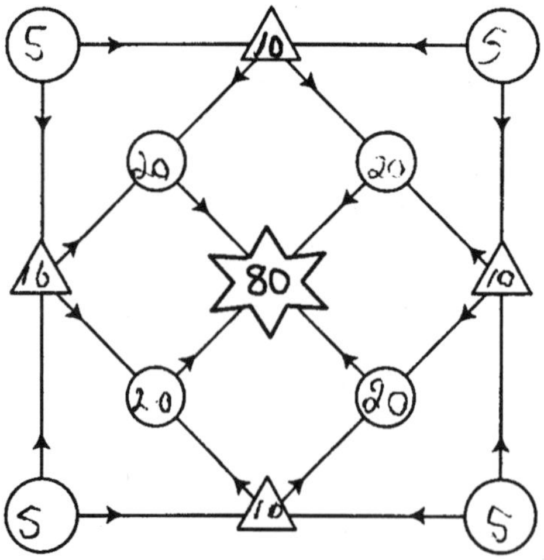

a. **Sum of the four beginning numbers** 20

b. **Star number** 80

Student D

7. Fill in this table by recording the sum of the four beginning numbers and the star number for each star number square. The first one is done for you.

Sum of beginning numbers	Star number
10	40
11	44
12	48
13	52

8. What rule would help you predict the star number if you know the sum of the four beginning numbers? I think the rule is 10, 11, 12, 13, x4. Like 10x4 is 40, 11x4 is 44, 12x4 is 48 and goes on and on, in the 40's It goes by 4 Like 40, 44, 48, 52.

Student D

9. Use the table to predict the star number when the sum of the beginning numbers is 20.

If I started with 20 the star number would be 80.

10. What do you think the beginning numbers can be when the star number is 48? I Think the Four numbers that add up to 12 is 3,3,3,3

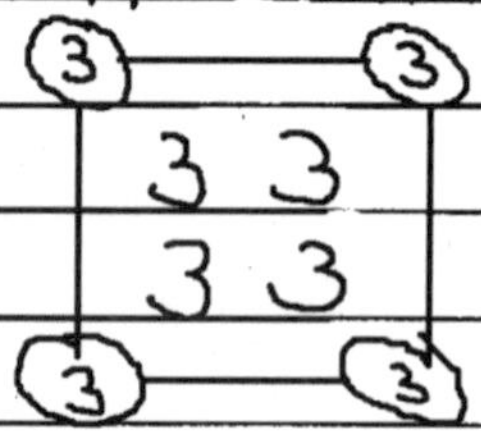

3 3
3 3

11. What other sets of numbers could be the beginning numbers for a star number of 48? How do you know?

I think another way is 4,3,2,3.

Task 4

Bolts and Nuts

Overview

Take pairs of measurements in a practical situation.

Make simple ratio calculations.

Long Task

Task Description

Students estimate, measure, and calculate the number of turns made to the nut and the distance it moves. Calculations bring in simple ideas of ratio, which pupils of this age will normally find quite challenging.

Assumed Mathematical Background

Students should have met the idea of ratio in a practical context. They should be familiar with measurements in millimeters and meters.

Core Elements of Performance

- take pairs of measurements in a practical situation
- make simple ratio calculations

Circumstances

Grouping:	Students may discuss the task in pairs, but each student should complete an individual written response.
Materials:	calculator, ruler, nut, and bolt The nut and bolt should not be connected together. The bolt should be at least 5 cm long and no more than 9 cm long. To help students count the turns, make a small mark with white correction fluid or a black permanent marker on one face of the nut and on one side of the head of the bolt.
Estimated time:	45 minutes

Name Date

Bolts and Nuts

This problem gives you the chance to

- *make measurements in a practical situation*
- *make simple ratio calculations*

Work with a partner on this problem.

You should have a nut, a bolt, a ruler, and a calculator.

Fix the nut onto the bolt.

Turn the nut so that it moves along the bolt.

1. How many complete turns will move the nut 25 mm along the bolt?

   First make a guess.

   Your guess: _________ turns move the nut 25 mm.

   Your partner's guess: _________ turns move the nut 25 mm.

   Talk to your partner about how you will answer this question. You will need to measure 25 mm with the ruler. There are some marks on the nut and bolt that may help you count the turns.

   Your answer: _________ turns move the nut 25 mm.

   Your partner's answer: _________ turns move the nut 25 mm.

   Do you agree? If not, who has the right answer?

© The Regents of the University of California

Name Date

Now try to answer the following questions *without touching the nut and bolt.*

2. How many times would you have to turn the nut to move it 50 mm? ______

3. Complete this table.

Distance the Nut Moves	Number of Turns	
25 mm		← Write in your answer from the previous page.
50 mm		← Write in your answer from the last question.
100 mm		← Figure this out.
200 mm		← Figure this out.
1 meter		← Figure this out.

To answer these you will need to imagine a bolt that is longer than the one you are using. The 1-meter bolt would be giant-sized because 1 meter = 1000 mm.

4. How far would the nut move if you turned it 50 complete turns? Explain how you figured this out. ______

______

______

______

© The Regents of the University of California

Task

A Sample Solution

The exact figures will of course depend on the type of nut and bolt. Here is a typical solution.

1.– 3.

Distance the Nut Moves	Number of Turns	
25 mm	About 20	← Answer to question 1.
50 mm	About 40	← Answer to question 2.
100 mm	About 80	← Figure this out.
200 mm	About 160	← Figure this out.
1 meter	About 800	← Figure this out.

4. If 1 meter or 1000 mm takes about 800 turns, then 125 mm takes about 100 turns. Therefore, 50 turns would move the nut about 62 mm.

Using this Task

Task

Arrange for students to work in pairs. Each student needs a calculator, a ruler, and a nut and bolt, which should not be connected together.

Demonstrate to the class how the nut can be attached to the bolt by twisting it on. Show how the nut can be moved along the bolt by turning it.

Ask students to try this. Encourage them to help each other. Offer as much help as you can to anyone who finds this difficult. Give them a little time to play, as this is important in order to understand the task.

Now demonstrate how to place the ruler next to the bolt and count the number of turns needed to move the nut a distance of 10 millimeters. It is important that they understand that "complete turn" means one complete revolution of the nut. Demonstrate this to them making use of the marks on the nut and bolt.

Now hand out the activity pages and read them with the class to make sure they understand what they have to do. Explain that everyone must hand in an individual response.

Task

Characterizing Performance

This section offers a characterization of student responses and provides indications of the ways in which the students were successful or unsuccessful in engaging with and completing the task. The descriptions are keyed to the *Core Elements of Performance.* Our global descriptions of student work range from "The student needs significant instruction" to "The student's work meets the essential demands of the task." Samples of student work that exemplify these descriptions of performance are included below, accompanied by commentary on central aspects of each student's response. These sample responses are *representative;* they may not mirror the global description of performance in all respects, being weaker in some and stronger in others.

The characterization of student responses for this task is based on these *Core Elements of Performance:*

1. Take pairs of measurements in a practical situation.
2. Make simple ratio calculations.

Descriptions of Student Work

The student needs significant instruction.

These papers show, at most, an attempt to make the estimates or measurements.

Student A

Only the estimates and measurements on the first page of the task have been attempted. This response shows no understanding of proportions.

The student needs some instruction.

An attempt has been made to make the estimates and measurements. The table has been at least partially completed, but there is a poor understanding of proportion.

Student B

The numbers in the table are 15, 30, 45, and 60; these are not in proportion. And, the student makes no attempt to answer question 4.

The student's work needs to be revised.

Measurements have been correctly made within the generous margin of error (± 4 turns). The table has been partially completed with most figures in proportion.

Student C

The numbers in the table show an understanding of proportion, although the figure for 1 meter is grossly incorrect.

The student's work meets the essential demands of the task.

Measurements have been correctly made within the generous margin of error (± 4 turns). The student may have indicated somewhere on the response that figures are "rough," "approximate," or "estimated." The student can handle simple ratio calculations as indicated either by the table completed with all figures in proportion, or by a consistent answer to the final question.

Student D

Although this response did not attempt question 4, the perfect completion of the table demonstrates a good understanding of proportion.

Student A

Bolts and Nuts

This problem gives you the chance to

- *make measurements in a practical situation*
- *make simple ratio calculations*

Work with a partner on this problem.

You should have a nut, a bolt, a ruler, and a calculator.

Fix the nut onto the bolt.

Turn the nut so that it moves along the bolt.

1. How many complete turns will move the nut 25 mm along the bolt?

   First make a guess.

   Your guess: ___15___ turns move the nut 25 mm.

   Your partner's guess : ___16___ turns move the nut 25 mm.

   Talk to your partner about how you will answer this question. You will need to measure 25 mm with the ruler. There are some marks on the nut and bolt that may help you count the turns.

   Your answer: ___16___ turns move the nut 25 mm.

   Your partner's answer: ___16___ turns move the nut 25 mm.

   Do you agree? If not, who has the right answer?

   We agree.

Student B

Fix the nut onto the bolt.

Turn the nut so that it moves along the bolt.

1. How many complete turns will move the nut 25 mm along the bolt?

   First make a guess.

   Your guess: 18 turns move the nut 25 mm.

   Your partner's guess : 18 turns move the nut 25 mm.

   Talk to your partner about how you will answer this question. You will need to measure 25 mm with the ruler. There are some marks on the nut and bolt that may help you count the turns.

   Your answer: 15 turns move the nut 25 mm.

   Your partner's answer: 15 turns move the nut 25 mm.

   Do you agree? If not, who has the right answer?

   We agree!

Now try to answer the following questions *without touching the nut and bolt.*

2. How many times would you have to turn the nut to move it 50 mm? ______

3. Complete this table.

Distance the Nut Moves	Number of Turns	
25 mm	15	← Write in your answer from the previous page.
50 mm	30	← Write in your answer from the last question.
100 mm	45	← Figure this out.
200 mm	60	← Figure this out.
1 meter	1000	← Figure this out.

To answer these you will need to imagine a bolt that is longer than the one you are using. The 1-meter bolt would be giant-sized because 1 meter = 1000 mm.

4. How far would the nut move if you turned it 50 complete turns? Explain how you figured this out. ______

______

Student C

Fix the nut onto the bolt.

Turn the nut so that it moves along the bolt.

1. How many complete turns will move the nut 25 mm along the bolt?

   First make a guess.

   Your guess: 18 turns move the nut 25 mm.

   Your partner's guess : 20 turns move the nut 25 mm.

   Talk to your partner about how you will answer this question. You will need to measure 25 mm with the ruler. There are some marks on the nut and bolt that may help you count the turns.

   Your answer: 16 turns move the nut 25 mm.

   Your partner's answer: 16 turns move the nut 25 mm.

   Do you agree? If not, who has the right answer?

   Yes we agree me and my partner got16.

Now try to answer the following questions *without touching the nut and bolt.*

2. How many times would you have to turn the nut to move it 50 mm? 32

3. Complete this table.

Distance the Nut Moves	Number of Turns	
25 mm	16	← Write in your answer from the previous page.
50 mm	32	← Write in your answer from the last question.
100 mm	64	← Figure this out.
200 mm	128	← Figure this out.
1 meter	32 768	← Figure this out.

Bolts and Nuts ■ Student Work

Student D

Fix the nut onto the bolt.

Turn the nut so that it moves along the bolt.

1. How many complete turns will move the nut 25 mm along the bolt?

   First make a guess.

   Your guess: 25 turns move the nut 25 mm.

   Your partner's guess : 15 turns move the nut 25 mm.

   Talk to your partner about how you will answer this question. You will need to measure 25 mm with the ruler. There are some marks on the nut and bolt that may help you count the turns.

   Your answer: 18 turns move the nut 25 mm.

   Your partner's answer: 18 turns move the nut 25 mm.

   Do you agree? If not, who has the right answer?

   yes we do we even double checked it!

Now try to answer the following questions *without touching the nut and bolt.*

2. How many times would you have to turn the nut to move it 50 mm? about 36

3. Complete this table.

Distance the Nut Moves	Number of Turns	
25 mm	18	← Write in your answer from the previous page.
50 mm	36	← Write in your answer from the last question.
100 mm	72	← Figure this out.
200 mm	144	← Figure this out.
1 meter	720	← Figure this out.

Overview

Apply mathematics to real-life situations.

Organize data.

Create a graph or table for a purpose.

Interpret data to solve a problem.

Field Trip

Long Task

Task Description

In this task, students use information from three different sources to produce a graph or table that clearly integrates and shows the information from all three sources in one display. They write a note interpreting the graph or table they constructed and make a recommendation based on their interpretation.

Assumed Mathematical Background

This task assumes that students have some prior experience organizing data into graphs and/or tables and in interpreting and using data to solve problems.

Core Elements of Performance

- organize data from a variety of sources
- make a graph
- use data to solve a real-life problem

Circumstances

Grouping:	Students may discuss the task in pairs, but each student should complete an individual written response.
Materials:	calculators
Estimated time:	45 minutes

Name Date

Field Trip

This problem gives you the chance to

- *use data from three classrooms to make a graph*
- *use the graph to decide where the fourth grade will go on its field trip*
- *figure out the cost of the field trip for each student*

The fourth-grade classes at Sunset Elementary School are going on a field trip together. They have a choice of three places to go (Discovery World, Modern Art Museum, or SeaTown). The students in each class took a vote to see which place most of them wanted to go. They agreed that they would go to the place that got the most votes. Here are the voting results from the three classes.

Mr. Metzler's Class

Field Trip Survey

Discovery World	*𝍸 𝍸 𝍸 III*
Modern Art Museum	*𝍸 II*
SeaTown	*𝍸 III*

Ms. Johnson's Class

Results of the vote for where the fourth grade should go on our field trip.

SeaTown	11
Discovery World	9
Modern Art Museum	12

Teacher: Ms. Lydon
Grade: 4 Class List

Amanda	Discovery World
Anthony	Art Museum
Antwan	Art Museum
Blanche	SeaTown
Brandon	Discovery World
Christine	SeaTown
Dannor	Art Museum
Dominic	Discovery World
Fiona	SeaTown
Gilbert	Discovery World
Gillian	Art Museum
Harriett	Discovery World
Hazel	SeaTown
Irene	SeaTown
Jonas	SeaTown
Jonibel	Discovery World
Joseph	Art Museum
Kenny	Discovery World
Kentasha	Art Museum
Kevin	Discovery World
Latasha	Discovery World
Maurice	SeaTown
Michael	Discovery World
Mindy	Art Museum
Nancy	SeaTown
Sabrina	Discovery World
Samantha	Discovery World
Satpreet	SeaTown
Suman	Discovery World
ToniAnn	Discovery World
Walter	Art Museum
Yolanda	Discovery World

© The Regents of the University of California

Name Date

The class secretary called all three places and asked them to send him information about admission prices. Here is what he found out.

Modern Art Museum

Group Rates for Schools:

Children to age 12	$2.00
Young people to age 18	$2.50
Chaperones	$3.00

Discovery World Admission Prices

Special Group Rates for Schools:

One Class — $60.00

Adult Chaperones — Free

The class secretary also checked on bus prices. This is what he found out.

School Buses for Field Trips

One bus can take only one class. Buses can be rented for the whole day for $50.00.

© The Regents of the University of California

Name Date

The class secretary wants your help. Here is a note from him telling you what he would like you to do.

> Dear Helper,
>
> I need to let all the fourth graders know the results of the class votes. I also need to write a letter to their parents asking them to send the money for the field trip. Please help me do two things.
>
> 1. Use the results from the votes in all three classes to make a graph or table that shows how many fourth graders voted to go to each place. The graph or table you make should be clear and easy to understand.
> 2. Then figure out how much the whole trip will cost and how much each student will need to pay. We need to pay for all the buses and the admission prices. Teachers and adult chaperones are not going to help us pay for the trip, but there must be 2 adult chaperones with each class. Every student will pay an equal share of the total. Write me a note explaining how much each student will need to pay. Be sure you explain how you figured it out, so I will know what to tell parents in the letter I write to them.
>
> Thanks,
> Class Secretary

You can put your graph or table on the next page, and your note to the class secretary can be written on the page after that.

© The Regents of the University of California

Name Date

You may create your graph or table on this grid.

© The Regents of the University of California

Name Date

You may write your note to the class secretary on this page.

© The Regents of the University of California

A Sample Solution

Task 5

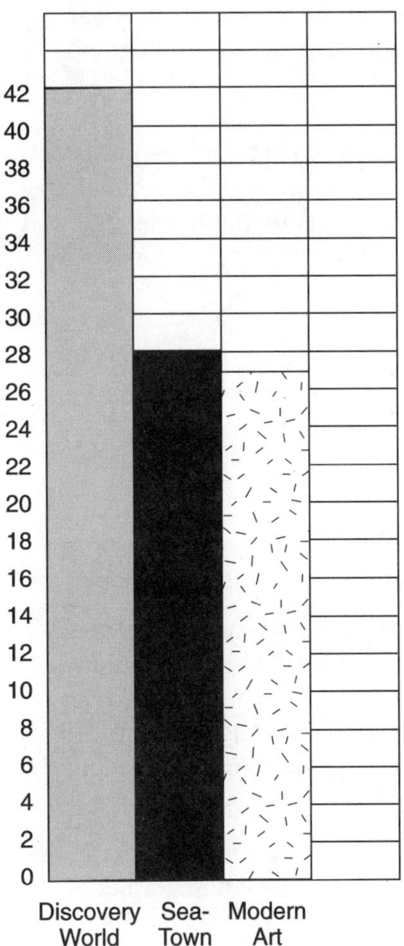

Any graph that shows that 42 students want to go to Discovery World, 28 want to go to SeaTown, and 27 want to go to the Modern Art Museum is acceptable. The graph scale may use increments other than 2. Most students will make a bar graph, but a correct circle graph is also acceptable.

The whole trip will cost \$330. Three classes need to pay \$60 each for admission to Discovery World ($3 \times \$60 = \$180$), and they will need three buses ($3 \times \$50 = \$150$). So the total cost is \$180 + \$150 = \$330. (Adult chaperones get in free at Discovery World, so there is no added cost for their admission.)

Since there are 97 students going, and teachers and chaperones are not helping to pay for the trip, the per pupil cost is \$3.40 ($330 \div 97 = 3.402\ldots$). If every student pays the \$3.40, they will need \$0.20 to reach the total cost. Some students may round up to \$3.41 in order to cover the full cost.

Task

Using this Task

This task can be done individually or in pairs. It is best done as an assessment after students have had an opportunity to read and interpret graphs and after they have made several graphs.

It is important to note that if students use calculators to figure the per pupil cost of the field trip, they will need to know what to do with the decimal that results from dividing 330 by 97. You may want to ensure that this assessment follows some instruction on interpreting the results of computations on a calculator.

Distribute calculators to students and read the task aloud. The task should take about 45 minutes to complete, but give students extra time if they are still working productively after 45 minutes.

Extensions

You may wish to extend this task by using some of the ideas that follow.

- Have students create graphs reflecting real issues and concerns in your classroom and school. Students may want to influence the school lunch program to add or eliminate particular choices. They could do a survey of the whole school, create graphs to show what students prefer, and use the graphs in combination with letters to the principal requesting certain changes.
- You could replicate this task on a real-life basis by allowing all the students at your grade level to vote on a field trip. The votes, graphing, and cost calculations could all be a part of planning a real field trip.

Characterizing Performance

Task

This section offers a characterization of student responses and provides indications of the ways in which the students were successful or unsuccessful in engaging with and completing the task. The descriptions are keyed to the *Core Elements of Performance.* Our global descriptions of student work range from "The student needs significant instruction" to "The student's work meets the essential demands of the task." Samples of student work that exemplify these descriptions of performance are included below, accompanied by commentary on central aspects of each student's response. These sample responses are *representative;* they may not mirror the global description of performance in all respects, being weaker in some and stronger in others.

The characterization of student responses for this task is based on these *Core Elements of Performance:*

1. Organize data from a variety of sources.
2. Make a graph.
3. Use data to solve a real-life problem.

Descriptions of Student Work

The student needs significant instruction.

These papers show evidence of an attempt to accomplish at least one of the core elements of performance. A response at this level of performance may show an attempt to organize the data from all three classes. The response may show an attempt at a graph or an attempt to figure out total cost and/or per pupil cost. All, or most, of these efforts are unsuccessful.

Student A

This student accurately counts how many students voted for each choice and how many students there are in total. There is, however, no graph showing this data. The response finds the total cost of admission to the Discovery World and total cost of the buses. The student does not find the total cost of the trip or the per pupil cost.

Task

The student needs some instruction.

These papers provide evidence of some success in at least one core element of performance. They may accurately figure out how many people voted to go to each place, but may not carry this information through to a successful graph. They may figure out the total cost of the trip without being able to calculate individual cost.

Student B

This response shows an attempt at a graph that "wraps around" the column for Discovery World and makes minor errors in counting the Modern Art Museum votes. Computations showing the total cost of the trip are on the response, but they are not labeled as such. The note does not state either total cost or per pupil cost.

Student C

This student numbers the boxes in the grid in columns for each choice. The Discovery World column "wraps around." There are errors in counting the votes for all three places. The number of students going on the trip is stated correctly, as is the total cost of the trip. There are no calculations shown to indicate how these numbers were computed. The price per pupil is incorrectly given as $1.50, with no explanation.

The student's work needs to be revised.

These responses provide evidence of success in at least two core elements of performance, and only minor mistakes in the other core element.

Typically, they figure out how many students voted for each field trip option and create a graph that is correct, or contains only very minor errors. They calculate the total price of the field trip, but may encounter some difficulty in dealing with the decimal.

Student D

This response shows a graph that has columns for each choice. Boxes in the columns contain Xs. The Discovery World column wraps around to the next column to allow for 42 votes in increments of one. The total cost of the field trip and the per pupil cost are correct.

The student's work meets the essential demands of the task.

Task

These responses successfully accomplish all three elements of performance. They may state that the cost per pupil is \$3.40 or \$3.41. In either case, they explain how much is needed to cover the total cost of the field trip. Graphs are complete and clearly labeled.

Student E

This response includes a graph that shows correctly how many voted to go to each place. The graph uses increments of two. The explanation of total cost and per pupil cost are clear and accurate. The response explains that when total cost was divided by 97, the result was 3.4020618. It explains why the decimal was rounded to \$3.41.

Student F

This response also shows a graph that uses increments of two and that is clearly labeled and correct, with the exception of a miscalculation for Sea-Town votes. The note explains total cost and per pupil cost clearly and accurately. The method used to figure per pupil cost was a guess-and-check approach.

Student A

You may create your graph or table on this grid.

Discovery World: 42
modern Art Museum: 27
Seatown: 28

142
27
28
97
Students

33
33
66

30
30
60

Discovery World: 15
Art Museum: 8
Seatown: 9

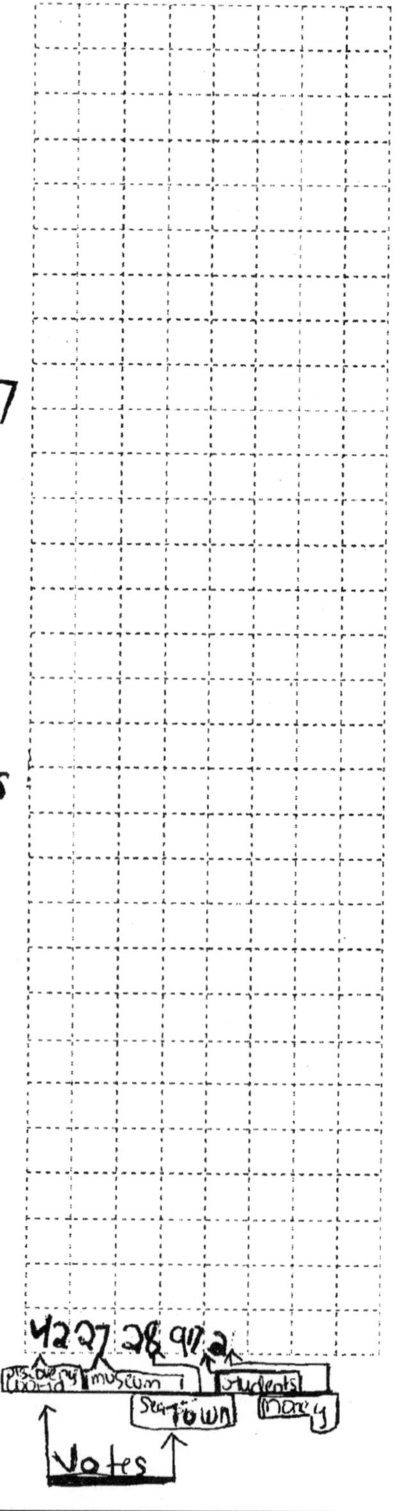

Student A

You may write your note to the class secretary on this page.

$2.940

It will cost (180) for the three buses.
It will cost (150) for Addmission.
I got (180) from the Three buses since each bus costs (60) dollars and 3x60=(180)
I got 150.

Student B

You may create your graph or table on this grid.

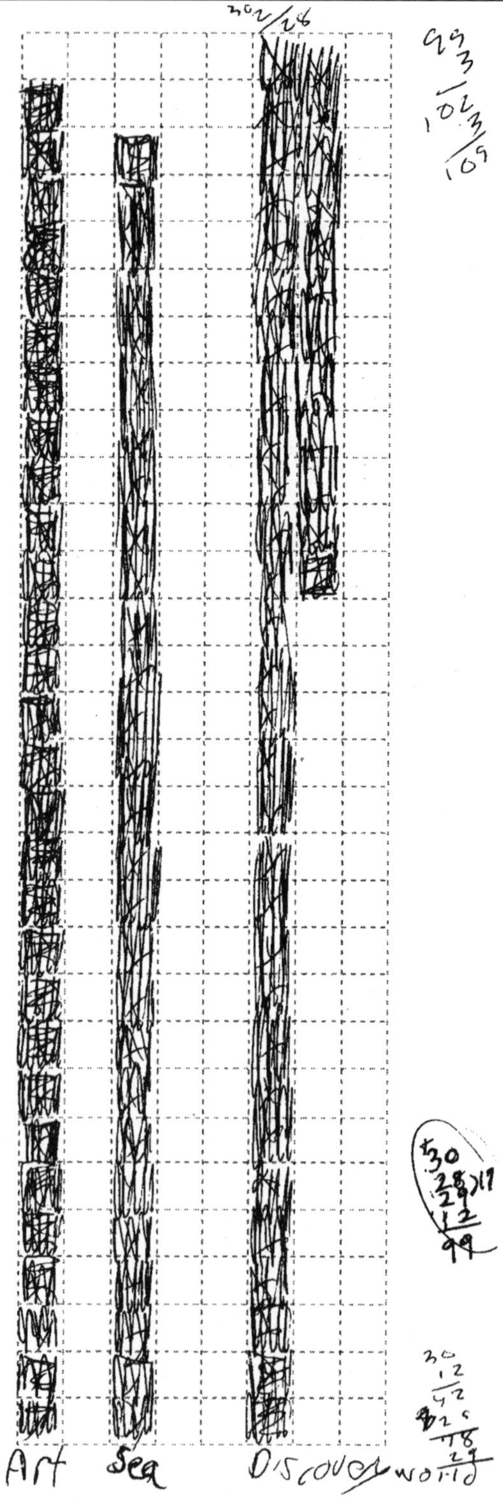

65
× 2
130

65
× 2
130

180
50
230

99)330
297
33

99
60
159

99 Kids
330 money

99
106
190

99
99
198

99
96
99
180

Art Sea Discovery world

Student B

You may write your note to the class secretary on this page.

Dear secratary,
I found out that everybody
have to bring because
3

50
x 150

60
x 3
180

99 kids 50
330

Student C

You may create your graph or table on this grid.

Disc	World	Sea Town	Museum
29			
28			
27			
26			
25			
24			
23			
22			
21			
20			
19		19	19
18		18	18
17		17	17
16		16	16
15		15	15
14		14	14
13		13	13
12	41	12	12
11	40	11	11
10	39	10	10
9	38	9	9
8	37	8	8
7	36	7	7
6	35	6	6
5	34	5	5
4	33	4	4
3	32	3	3
2	31	2	2
1	30	1	1
Disc	World	Sea Town	Madeth Museum
41		14	19

Student C

You may write your note to the class secretary on this page.

Dear 4th grade Secertary,

This is how I figured how many people are going were they are going and how much money it costs and how much money the kids should bring.
I figured how many people are going. I counted up how many people there were in each class then added them all up and got 97 kids. It will cost 330.00 all together Each kid should bring $1.50 for the bus and field trip. Most of the kids voted for the Discovery World so thats where they're going to have a good field trip

Student D

You may create your graph or table on this grid.

$150
180
$330

91)330

33
32
32
97

3.40

91)330

Student D

You may write your note to the class secretary on this page.

Dear Secretary,
It turns out that the place the fourth grade is going is Discovery World each person will pay $3.40 so we can cover the buses and the place. The way I figured it out by dividing 330 by 97 and got 3.40. I counted the students and got 97 and then I multiplied the cost of everything which was 50×3=150 and 60×3=180 and then I added it which made $330 as I have shown above. I have 77¢ left

Student E

You may create your graph or table on this grid.

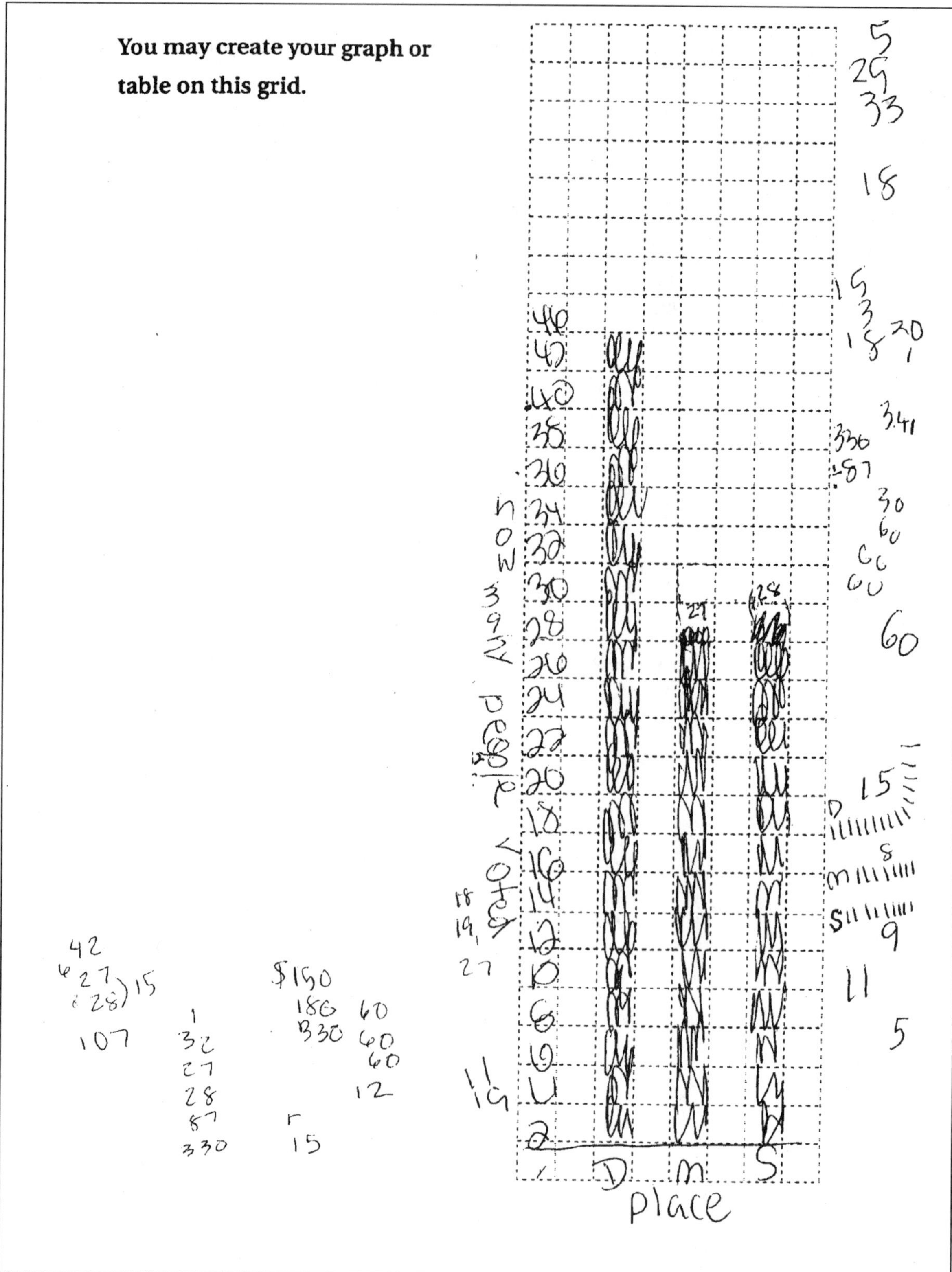

Student E

You may write your note to the class secretary on this page.

Dear Secretary,
I found out the fourth grade will goto Discovery World.
Each Student Will pay $3.41 because there are 3 classes Each class cost $50.00. 50x3 is 150. There are 3 buses each one cost $60. $60x3 is 180. 180 plus 150 is $330.00.
There are 97 students. 330 ÷ 97 is 3.4020618. The numers after the dot are cents. 40 are whole cents and 20618. is part of a penny. So you put in a penny to take care of it and it cost $3.41

Student F

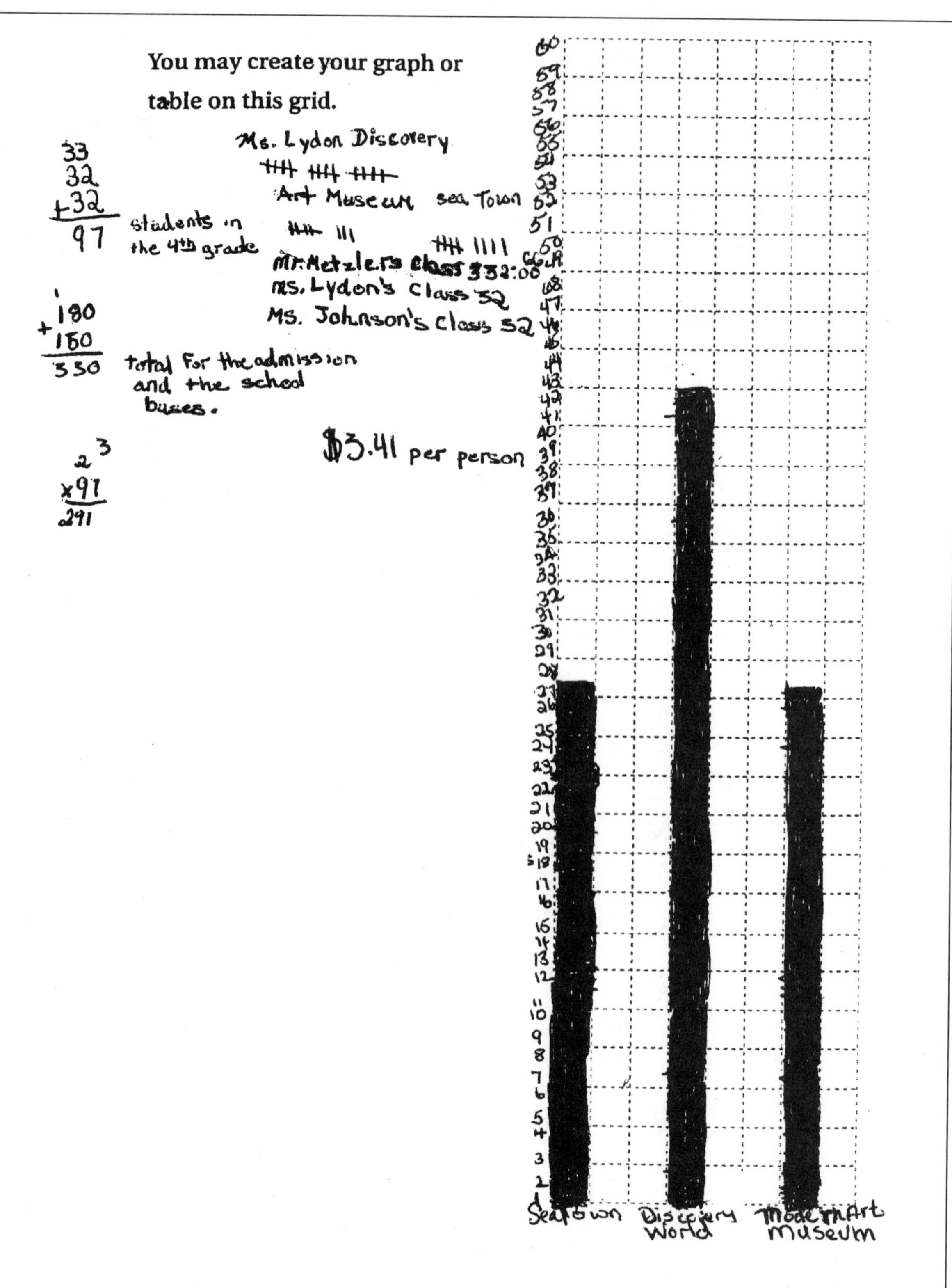
You may create your graph or
table on this grid.
33
32
+32
97 students in the 4th grade
Ms. Lydon Discovery
Art Museum
Sea Town
Mr. Metzler's class 33
Ms. Lydon's class 32
Ms. Johnson's class 32
180
+150
350 total for the admission and the school buses.
$3.41 per person
x97
291
Seatown
Discovery World
Modern Art Museum

Student F

You may write your note to the class secretary on this page.

Dear Secretary,

the vote for the best fieldtrip was, Sea World 27, Modern Art Museum 27, and Discovery World 42. The admission for Discovery World is $50 per class. There are 3 fourth grade classes to get in plus it is $180 dollars for 3 busses. $130 plus $160 equals $330. First I tried $3.50 It was about 9 dollars to much so I tried $3.40 It was 20 cents under $330 so I tried $3.41. It was 71 cents bigger than $330 but it wasn't lower so it was the right answer.

Task 6

Overview

Identify geometric shapes.

Use line symmetry to complete a design.

Describe the design.

Old Ruins

Long Task

Task Description

Students are asked to describe and draw designs made from simple geometric shapes.

Assumed Mathematical Background

It is assumed that students have prior experience using mathematical language to identify geometric shapes and an understanding of line symmetry.

Core Elements of Performance:

- identify geometric shapes within a design
- use line symmetry to complete a design
- describe the completed design using mathematical language

Circumstances

Grouping:	Students work to complete an individual written response.
Materials:	rulers and colored pencils
Estimated time:	45 minutes

Name Date

Old Ruins

This problem gives you the chance to

- *find geometric shapes that are part of a design*
- *complete a geometric design using line symmetry*
- *write a mathematical description of your design*

Here is a floor pattern found in some old ruins.

1. Below is Ted's unfinished description of this floor pattern. Help Ted by filling in the missing numbers.

   The 8-pointed star in the center of the design is made of __________ parallelograms.

   Touching the edges of the 8-pointed star are __________ triangles and __________ squares.

   Around the edges of this design are __________ more triangles and __________ more parallelograms. These parallelograms are bigger than those that meet at the center.

2. Color the floor pattern using the same color for shapes that are the same.

© The Regents of the University of California

Name Date

On this page and the next page are diagrams showing parts of two different floor designs.

The dotted lines show two lines of symmetry, but there may be others!

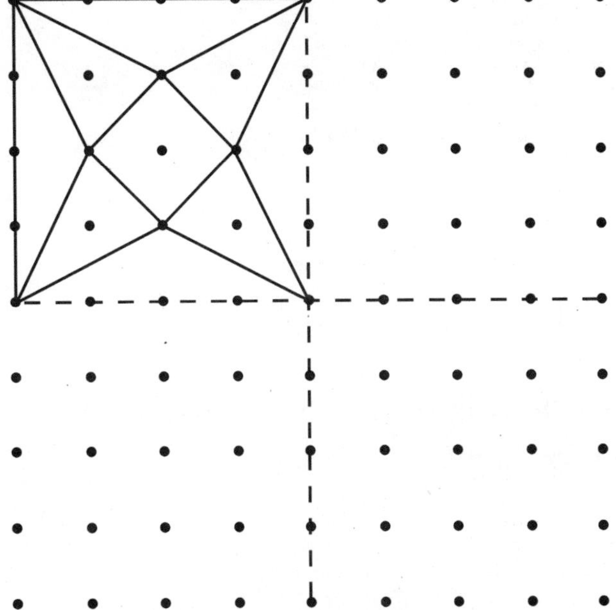

3. Complete this floor design on the grid provided and then describe the design in the space below. ______________________

© The Regents of the University of California

Name Date

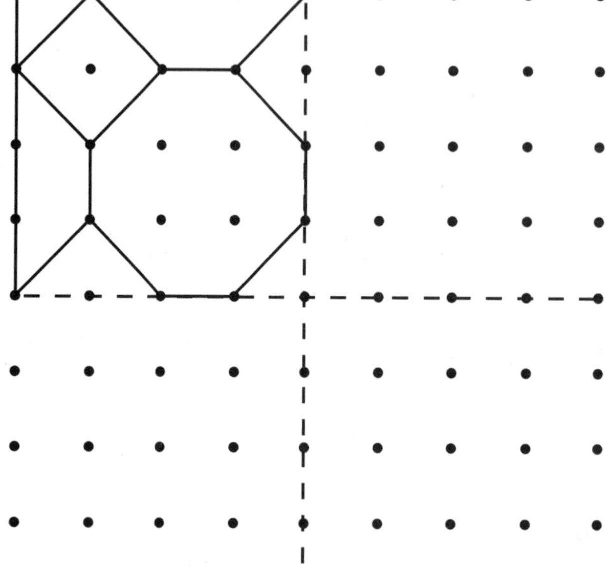

4. Complete this floor design on the grid provided and then describe the design in the space below. ______

______

______

______

______

______

______

© The Regents of the University of California

A Sample Solution

Task

1. The 8-pointed star in the center of the design is made of **8** parallelograms. Touching the edges of the 8-pointed star are **4** triangles and **4** squares. Around the edges of this design are **4** more triangles and **8** more parallelograms. These parallelograms are bigger than those that meet at the center.

2. Check that the diagram has been colored using the same color for shapes that are the same.

3.

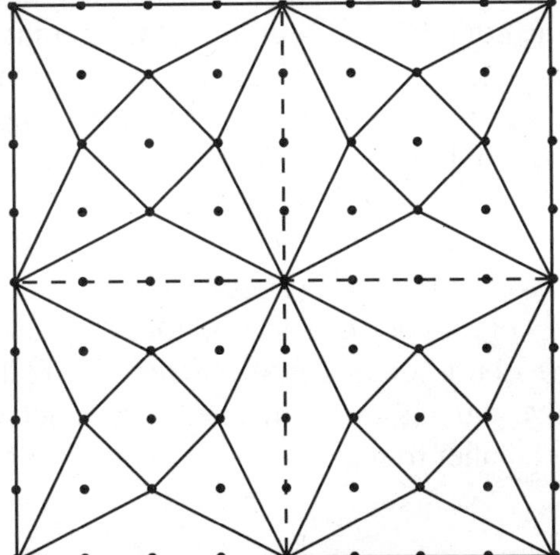

When the floor is completed, there are four parts that are identical to the original design piece. Each part is made up of a small square, fixed in a diagonal position, with four (isosceles) triangles on its sides. Around the edges are four (isosceles) triangles, different from the first ones. Altogether, there are four small squares, sixteen (isosceles) triangles of the first type, eight (isosceles) triangles of the second type, and four rhombuses: along each line of symmetry two isosceles triangles make a rhombus. However, if students consider the four rhombuses along the lines of symmetry to be eight triangles, this is also acceptable.

The original design has lines of symmetry through its center that are parallel to the dotted lines of symmetry for the whole floor. Consequently, each of the three parts that students draw should not be merely reflections, but actually identical to the design provided.

Task

4.

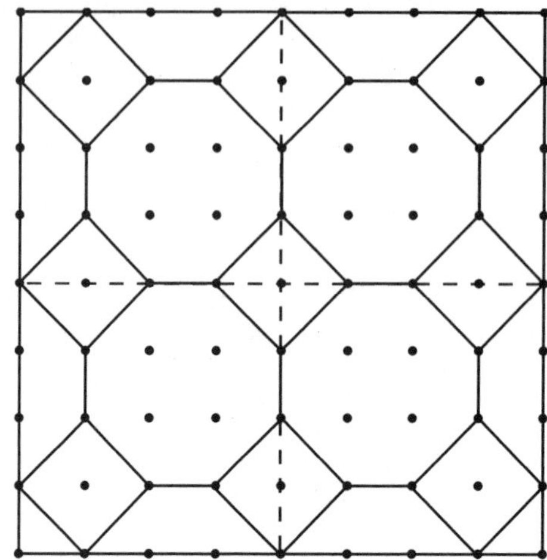

When the floor has been completed, there are four parts that are identical to the original design piece, but placed in four different orientations. Each part contains an octagon; a square fixed in a diagonal position, which touches an edge of the octagon; two trapezoids which each touch an edge of the octagon; and four right triangles.

Altogether the floor contains four octagons; eight trapezoids which each touch an edge of an octagon; nine squares each fixed in a diagonal position and touching the edge(s) of (an) octagon(s); and four right triangles each found touching the edge of the squares at the corners of the floor. If students consider the five squares along the lines of symmetry to be eight triangles and four smaller triangles, this is also acceptable.

Using this Task

Task

Read through the prompt with your students to ensure that they understand the task and answer any questions that arise. Be sure students understand that they need to complete each design and then describe it as precisely as they can.

Extensions

You may wish to extend this task by using some of the ideas that follow.

- Identify geometric shapes and lines of symmetry in floor coverings and wall tiles for kitchens and bathrooms.
- Explore symmetry patterns in wallpaper and fabric designs.
- Design a tile and use it to cover a surface.
- Draw a tile design using LOGO or another computer drawing program.
- Identify geometric shapes and lines of symmetry in quilt designs.

Characterizing Performance

Task

This section offers a characterization of student responses and provides indications of the ways in which the students were successful or unsuccessful in engaging with and completing the task. The descriptions are keyed to the *Core Elements of Performance.* Our global descriptions of student work range from "The student needs significant instruction" to "The student's work meets the essential demands of the task." Samples of student work that exemplify these descriptions of performance are included below, accompanied by commentary on central aspects of each student's response. These sample responses are *representative;* they may not mirror the global description of performance in all respects, being weaker in some and stronger in others.

The characterization of student responses for this task is based on these *Core Elements of Performance:*

1. Identify geometric shapes within a design.
2. Use line symmetry to complete a design.
3. Describe the completed design using mathematical language.

Descriptions of Student Work

The student needs significant instruction.

These papers show evidence of the first core element of performance. They identify geometric shapes such as triangles, squares, and parallelograms within a design.

Student A

This student has correctly identified shapes within the first design. He has also correctly completed the design in problem 3, but he has not provided a description.

The student needs some instruction.

These papers show evidence of success in the first element of performance and partial success in the second element of performance. They identify, and count correctly, geometric shapes such as triangles, squares, and parallelograms within a design. They show partial understanding of line symmetry by completing at least one section of one of the floor designs correctly.

Student B

This response correctly identifies shapes within the design in problem 1. It correctly completes the design in problem 3, but the design in problem 4 is incorrect. There are descriptions of both designs, but neither contains any mathematical language relating to the shapes within the designs.

The student's work needs to be revised.

These papers show some evidence of all three core elements of performance. However, the work on core elements two or three may be incomplete or may contain errors. These responses identify and correctly count the geometric shapes within a design. They show an understanding of line symmetry by completing at least one of the two floor designs correctly. One of the two diagrams may be incompletely or incorrectly drawn. Frequently, each section of the final floor design is drawn with the same orientation as the section provided, rather than being a reflection of it. They use mathematical language to describe the completed designs. These descriptions, however, may be incomplete.

Student C

This student correctly identifies the shapes in problem 1. In problem 3, the design is completed correctly and mathematical shapes are identified. Although the design in problem 4 has not been completed, some understanding of line symmetry is displayed.

The student's work meets the essential demands of the task.

These papers show evidence of success in all three core elements of performance. They identify and correctly count the geometric shapes within a design. They complete both floor patterns, however, the completed designs may contain minor errors. They use mathematical language to describe the completed designs. The descriptions state the types of shapes and tell how many are in each design.

Student D

This response identifies shapes within the design in problem 1 and correctly completes the designs in problems 3 and 4. It contains written descriptions of the completed designs using some mathematical language. The use of "oxegons" rather than octagons does not reduce the value of this paper.

Student A

Old Ruins

This problem gives you the chance to

- *find geometric shapes that are part of a design*
- *complete a geometric design using line symmetry*
- *write a mathematical description of your design*

Here is a floor pattern found in some old ruins.

1. Below is Ted's unfinished description of this floor pattern. Help Ted by filling in the missing numbers.

   The 8-pointed star in the center of the design is made of __8__ *parallelograms.*

   Touching the edges of the 8-pointed star are __4__ *triangles and* __4__ *squares.*

   Around the edges of this design are __4__ *more triangles and* __8__ *more parallelograms. These parallelograms are bigger than those that meet at the center.*

2. Color the floor pattern at the top of this page, using the same color for shapes that are the same.

Student A

On this page and the next page are diagrams showing parts of two different floor designs.

The dotted lines show two lines of symmetry, but there may be others!

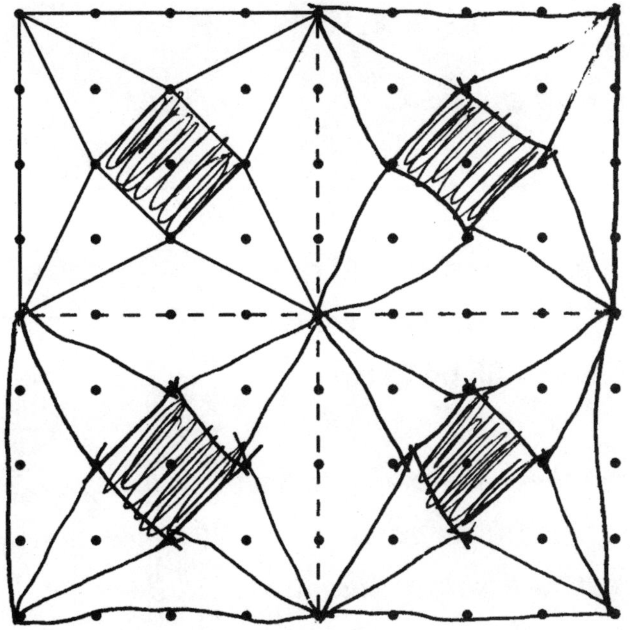

3. Complete this floor design on the grid provided and then describe the design in the space below.

Student B

Old Ruins

This problem gives you the chance to

- *find geometric shapes that are part of a design*
- *complete a geometric design using line symmetry*
- *write a mathematical description of your design*

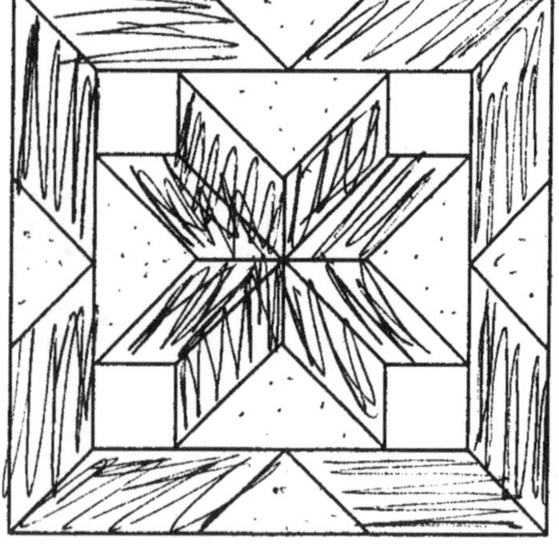

Here is a floor pattern found in some old ruins.

1. Below is Ted's unfinished description of this floor pattern. Help Ted by filling in the missing numbers.

   The 8-pointed star in the center of the design is made of __8__ *parallelograms.*

   Touching the edges of the 8-pointed star are __4__ *triangles and* __4__ *squares.*

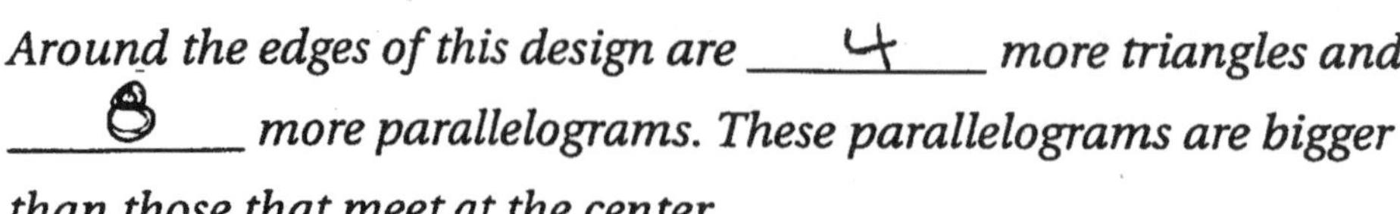

   Around the edges of this design are __4__ *more triangles and* __8__ *more parallelograms. These parallelograms are bigger than those that meet at the center.*

2. Color the floor pattern at the top of this page, using the same color for shapes that are the same.

Student B

On this page and the next page are diagrams showing parts of two different floor designs.

The dotted lines show two lines of symmetry, but there may be others!

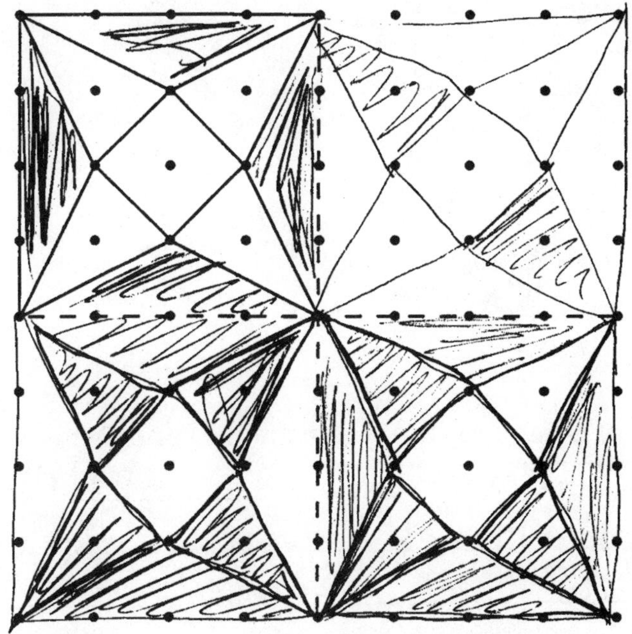

3. Complete this floor design on the grid provided and then describe the design in the space below. I describe thispicture by telling the person that diangnally, on the edges are, red going right Going left, the edges are green there different colors exept for the middle of each star.

Student B

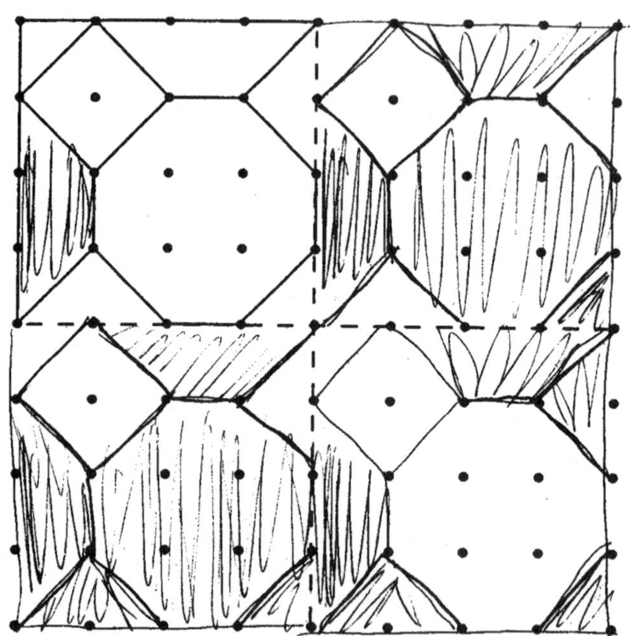

4. Complete this floor design on the grid provided and then describe the design in the space below. I would describe this picture by saying in each square all the colors are the same and have bright colors in them. All these colors are unusual.

Student C

Old Ruins

This problem gives you the chance to

- *find geometric shapes that are part of a design*
- *complete a geometric design using line symmetry*
- *write a mathematical description of your design*

Here is a floor pattern found in some old ruins.

1. Below is Ted's unfinished description of this floor pattern. Help Ted by filling in the missing numbers.

   The 8-pointed star in the center of the design is made of __8__ *parallelograms.*

   Touching the edges of the 8-pointed star are __4__ *triangles and* __4__ *squares.*

   Around the edges of this design are __4__ *more triangles and* __8__ *more parallelograms. These parallelograms are bigger than those that meet at the center.*

2. Color the floor pattern at the top of this page, using the same color for shapes that are the same.

Student C

On this page and the next page are diagrams showing parts of two different floor designs.

The dotted lines show two lines of symmetry, but there may be others!

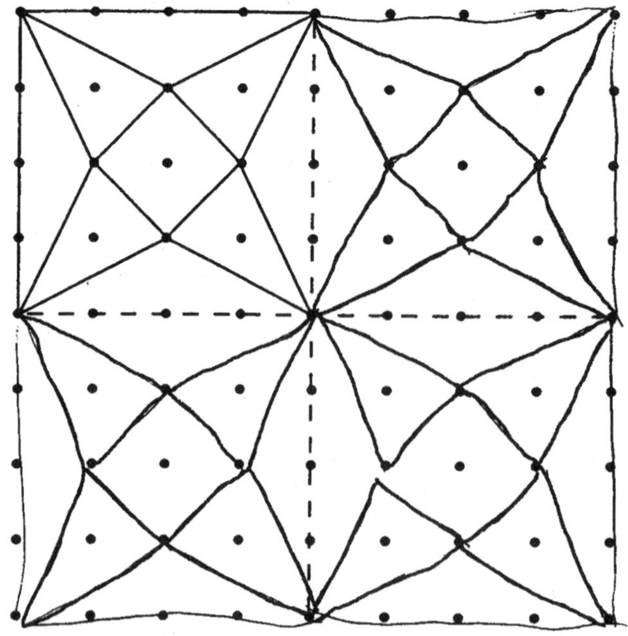

3. Complete this floor design on the grid provided and then describe the design in the space below.

There were 4 squarse.
There were 16 Tryangle
there were 4 dlments
There were 4 stars.
Ther were 16 half aimends

Student C

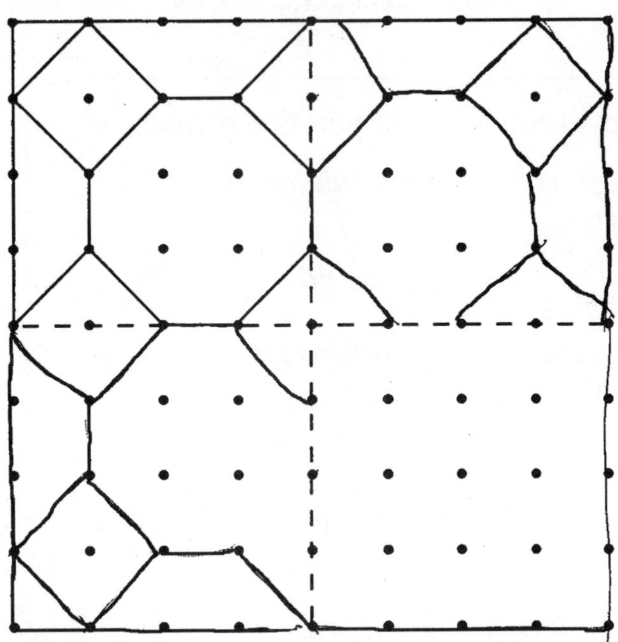

4. Complete this floor design on the grid provided and then describe the design in the space below. ____________________

Student D

Old Ruins

This problem gives you the chance to

- *find geometric shapes that are part of a design*
- *complete a geometric design using line symmetry*
- *write a mathematical description of your design*

Here is a floor pattern found in some old ruins.

1. Below is Ted's unfinished description of this floor pattern. Help Ted by filling in the missing numbers.

   The 8-pointed star in the center of the design is made of __8__ *parallelograms.*

   Touching the edges of the 8-pointed star are __4__ *triangles and* __4__ *squares.*

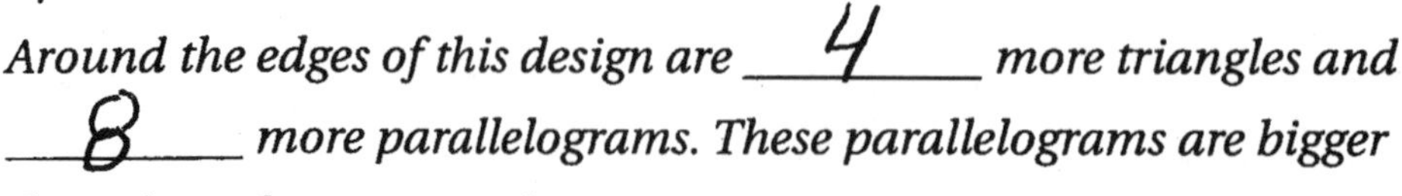

   Around the edges of this design are __4__ *more triangles and* __8__ *more parallelograms. These parallelograms are bigger than those that meet at the center.*

2. Color the floor pattern at the top of this page, using the same color for shapes that are the same.

Student D

On this page and the next page are diagrams showing parts of two different floor designs.

The dotted lines show two lines of symmetry, but there may be others!

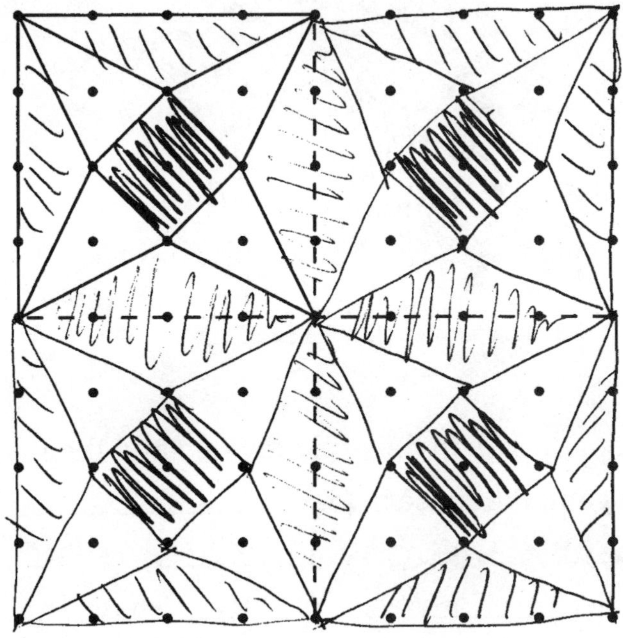

3. Complete this floor design on the grid provided and then describe the design in the space below. In the middle of this patten there are 4 triangles and 4 lardge dimonds they all meet at 45° then it has one square witch is attached to one of the points of two dimonds then there is another triangle witch goes to one point of the box that it's in. In the edge of the box next to one triangle is another lardger triangle

Student D

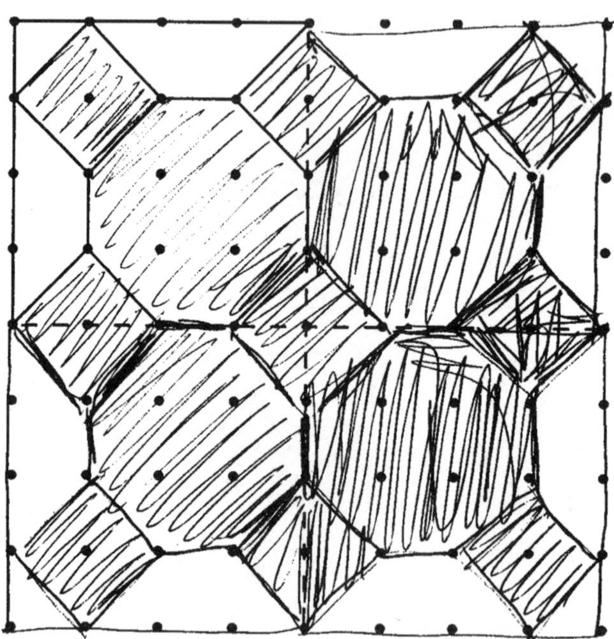

4. Complete this floor design on the grid provided and then describe the design in the space below.

The one at the top of this page has one little square in the middle of the page then it is surounded by four Oxegons then there is three squares on the top three down both sides and three down at the bottom. Then it's in a box.

Task 7

Overview

Use the concept of rate to solve a problem.

Calculate and compare number and quantity accurately.

Verify mathematical reasoning.

Millie's Business

Short Task

Task Description

In this task students are asked to figure out which of two kinds of supplies would be used up first, given the amount of each supply on hand and the amount of each supply used each day. Students are asked to explain or show how they figured out their answers.

Assumed Mathematical Background

Students should have prior experience using problem solving strategies, such as breaking a problem down into smaller parts and sketching or acting out part of the situation. They should also have experience with using everyday fractional amounts such as $\frac{1}{2}$ and $\frac{1}{4}$ to answer mathematical questions and solve problems.

Core Elements of Performance

- note the important comparative features of the situation
- use rate (for example, gallons per day and pints per day) to compare number and quantity accurately
- verify one's own mathematical reasoning

Circumstances

Grouping:	Students work to complete an individual written response.
Materials:	No special materials are needed for this task.
Estimated time:	15 minutes

Name Date

Millie's Business

This problem gives you the chance to

- *use measures of gallons, pints, and days to solve a mathematical problem*
- *work accurately with numbers (including fractions)*
- *explain the math in your answer*

Millie has a lawn-mowing business. She uses $\frac{1}{4}$ pint of oil each day and $1\frac{1}{2}$ gallons of gasoline each day. She has these supplies on her shelf:

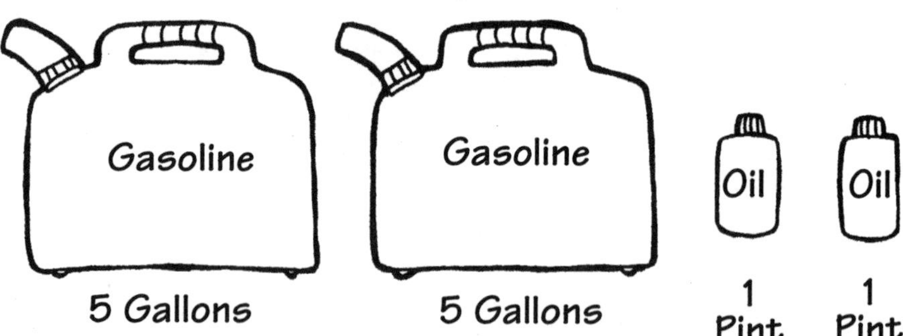

Which will she run out of first, gasoline or oil? Show or explain how you know your answer is right. ______________________

____

____

____

____

© The Regents of the University of California

A Sample Solution

Task

"One fourth" means "one out of four," so it takes four days to use up $\frac{4}{4}$, or one pint of oil. It takes four more days to use up the second pint, for a total of eight complete days.

Every two days 3 gallons of gasoline are used ($1\frac{1}{2} + 1\frac{1}{2} = 3$), so at the end of six days 9 gallons of gasoline will have been used. Millie will run out of gasoline partway through the seventh day, more than a day before she runs out of oil.
OR
Look at the table that shows how much of each kind of supply is used up at the end of each day:

Number of days	Gallons used: Gasoline	Pints used: Oil
1	$1\frac{1}{2}$	$\frac{1}{4}$
2	3	$\frac{1}{2}$
3	$4\frac{1}{2}$	$\frac{3}{4}$
4	6	1
5	$7\frac{1}{2}$	$1\frac{1}{4}$
6	9	$1\frac{1}{2}$
7	run out	$1\frac{3}{4}$
8		2

During the seventh day, the gasoline runs out. The oil lasts until the end of the eighth day.
OR
Since half of the oil on hand (1 pint) lasts four full days, find the answer by figuring out whether or not at least half of the gasoline (5 gallons) will be used up at the end of four days. Whichever supply is first to be half used up will be first to be entirely used up. Three gallons of gasoline is used up in two days, so 6 gallons (more than half of the 10 gallons on hand) will be used up at the end of four days. The gasoline will therefore be used up first.

Task

Using this Task

Some students who do well noting the important comparative features (there is less oil on hand, more gasoline gets used each day), stop there and make a guess, rather than going on to figure out the solution. Therefore, when presenting this task, consider stressing the importance of "showing or explaining how you know your answer is right" before students begin to work on the task.

Reviewing students' successful solutions is particularly useful with this task, since there is one answer, but a number of ways to solve it. Students whose solutions were partially successful—in that they knew what the important comparative features were, but didn't manage them successfully to solve the problem—can benefit from seeing some of the ways other students broke the task into smaller, more manageable pieces.

Extensions

You may wish to extend this task by using some of the ideas that follow.

- Keeping the rates of usage the same, ask students to figure out the amount of oil Millie would use in the time she uses 15 gallons of gasoline.
- To get at a different mathematical idea (least common multiple), have students work on another variation: Millie starts Day 1 by opening a fresh 5-gallon can of gasoline and a fresh pint of oil at the same time. But of course she won't finish that 5-gallon can at the same time as the pint of oil; that will hardly ever happen. Every once in a while, though, it will come out even—she will finish both kinds of container at the same time. Which day will be the first time that she finishes a can of gasoline and a pint of oil at the same time?

Characterizing Performance

Task

This section offers a characterization of student responses and provides indications of the ways in which the students were successful or unsuccessful in engaging with and completing the task. The descriptions are keyed to the *Core Elements of Performance.* Our global descriptions of student work range from "The student needs significant instruction" to "The student's work meets the essential demands of the task." Samples of student work that exemplify these descriptions of performance are included below, accompanied by commentary on central aspects of each student's response. These sample responses are *representative;* they may not mirror the global description of performance in all respects, being weaker in some and stronger in others.

The characterization of student responses for this task is based on these *Core Elements of Performance:*

1. Note the important comparative features of the situation.
2. Use rate (for example, gallons per day and pints per day) to compare number and quantity accurately.
3. Verify one's own mathematical reasoning.

Descriptions of Student Work

The student needs significant instruction.

These papers show, at most, evidence of understanding the important comparative features of the situation. Typically these papers will make a statement ("gasoline will run out first" or "oil will run out first"), and note a comparative feature in support of the statement (for example, "she uses more gasoline each day" or "there is less oil on hand"). These papers may note additional features that do not support the statement.

Student A

This student states that the gasoline will run out first and supports this statement by noting an important comparative feature of the situation: that a greater quantity of gasoline is used each day.

Task

Student B

This response states that the oil will run out first and supports this statement by noting an important comparative feature of the situation: that there is less oil than gasoline on hand to use.

The student needs some instruction.

These papers may state that the gasoline or the oil will run out first. These papers make a quantitative statement that is either inaccurate or partly inaccurate (for example, it may be accurate for one gallon of gasoline, but overlooks the second gallon of gasoline).

Student C

This response attempts to show that the oil will be used up in four days. This shows an understanding of the day-by-day use of oil for one pint. The second pint is not figured, nor is the use of gasoline.

The student's work needs to be revised.

These papers may state that the gasoline or the oil will run out first. These papers show evidence of understanding more than one comparative feature of the situation and show a workable, though incorrectly carried out, strategy for determining which item will be used up first.

Typically, these papers show that the student figured out the amount of time the gasoline (or oil) would last, but the other supply, though attempted, is figured inaccurately.

Student D

This response states that the oil will run out first. The number of days the gasoline will last is shown (the drawings of "gallons" are each split into five parts and the statement, "Only can use 7 times" is accurate). The number of days the oil will last is shown, but it is incorrect and seems to be based on the assumption that a whole pint of oil is used each day.

Student E

This response states that the oil will run out first. The number of days the oil will last is shown and is accurate. The number of days the gasoline will last is shown, but is incorrect and seems to be based on the assumption that one gallon, rather than $1\frac{1}{2}$ gallons, is used each day.

The student's work meets the essential demands of the task.

Task

These papers state that the gasoline will run out first. These papers show evidence of understanding both the comparative and quantitative features of the situation; these papers also show the mathematically sound strategy the student used to arrive at a solution. It is not necessary that the response show exactly when the gasoline runs out, only that it demonstrate mathematically that the gasoline will run out first.

Student F

This response states that the gasoline will run out first. The response states, accurately, that in six days 9 gallons of gasoline are used and in eight days 2 pints of oil are used.

Student A

Millie's Business

This problem gives you the chance to

- *use measures of gallons, pints, and days to solve a mathematical problem*
- *work accurately with numbers (including fractions)*
- *explain the math in your answer*

Millie has a lawn-mowing business. She uses $\frac{1}{4}$ pint of oil each day and $1\frac{1}{2}$ gallons of gasoline each day. She has these supplies on her shelf:

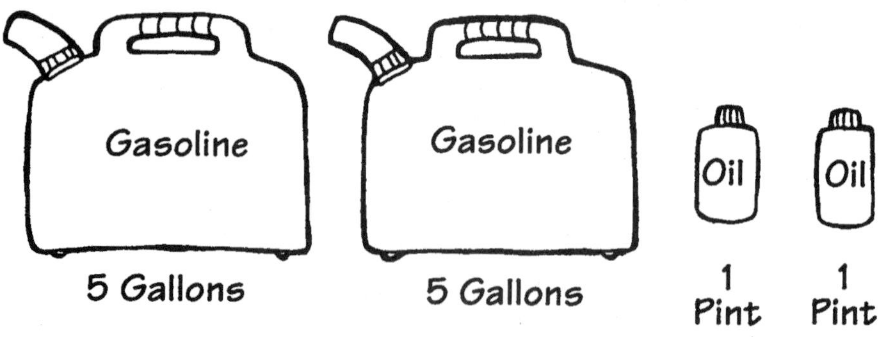

Which will she run out of first, gasoline or oil? Show or explain how you know your answer is right. I think she would run out of gasoline faster because she uses more of it each day.

Student B

Millie's Business

This problem gives you the chance to

- *use measures of gallons, pints, and days to solve a mathematical problem*
- *work accurately with numbers (including fractions)*
- *explain the math in your answer*

Millie has a lawn-mowing business. She uses $\frac{1}{4}$ pint of oil each day and $1\frac{1}{2}$ gallons of gasoline each day. She has these supplies on her shelf:

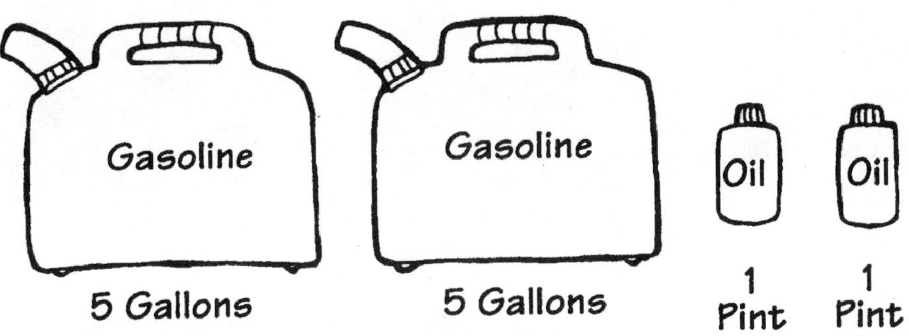

Which will she run out of first, gasoline or oil? Show or explain how you know your answer is right.

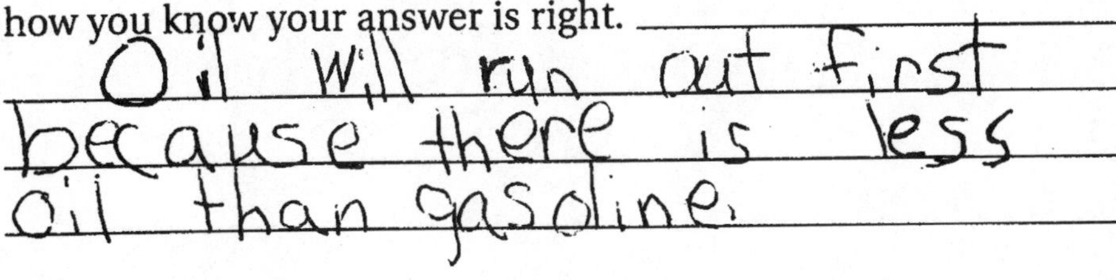

Student C

Millie's Business

This problem gives you the chance to

- *use measures of gallons, pints, and days to solve a mathematical problem*
- *work accurately with numbers (including fractions)*
- *explain the math in your answer*

Millie has a lawn-mowing business. She uses $\frac{1}{4}$ pint of oil each day and $1\frac{1}{2}$ gallons of gasoline each day. She has these supplies on her shelf:

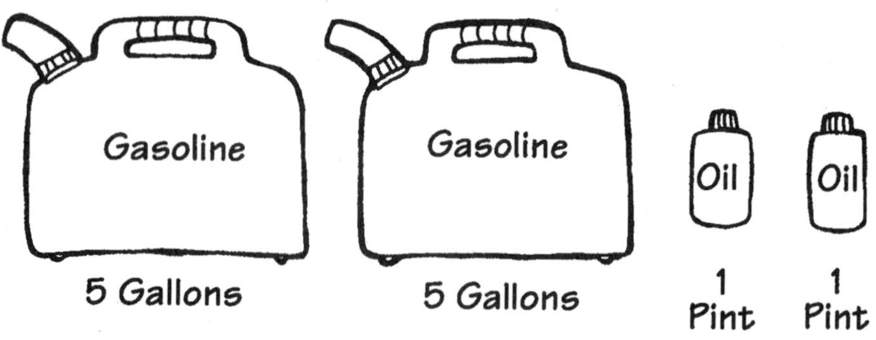

Which will she run out of first, gasoline or oil? Show or explain how you know your answer is right. 1 pint because 2 pint sudtract 4 each day you would have no more in four days.

Student D

Millie's Business

This problem gives you the chance to

- *use measures of gallons, pints, and days to solve a mathematical problem*
- *work accurately with numbers (including fractions)*
- *explain the math in your answer*

Millie has a lawn-mowing business. She uses $\frac{1}{4}$ pint of oil each day and $1\frac{1}{2}$ gallons of gasoline each day. She has these supplies on her shelf:

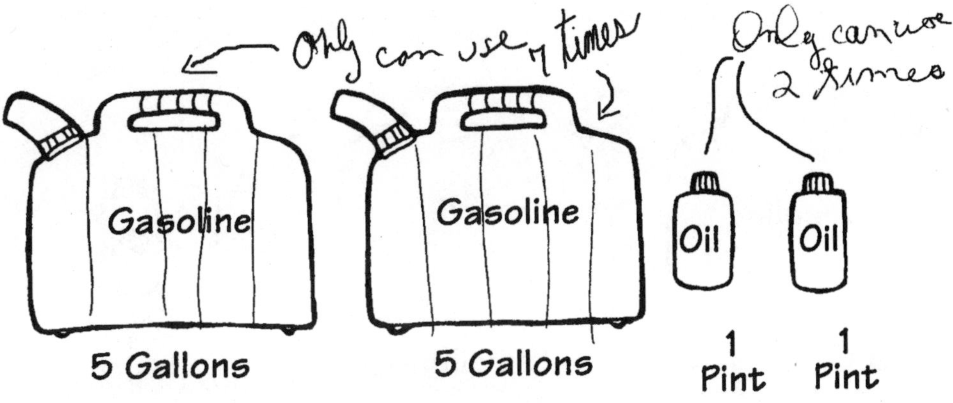

Which will she run out of first, gasoline or oil? Show or explain how you know your answer is right. Millie is going to run out of oil first becaus she has more gasoline than oil but she used a lot of oil in the pints.

Student E

Millie's Business

This problem gives you the chance to

- *use measures of gallons, pints, and days to solve a mathematical problem*
- *work accurately with numbers (including fractions)*
- *explain the math in your answer*

Millie has a lawn-mowing business. She uses $\frac{1}{4}$ pint of oil each day and $1\frac{1}{2}$ gallons of gasoline each day. She has these supplies on her shelf:

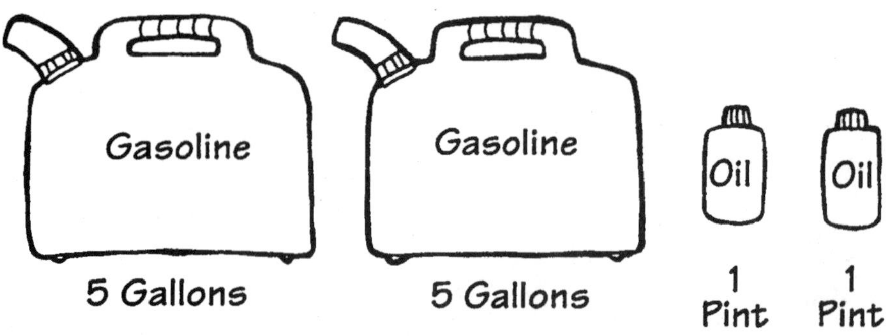

Which will she run out of first, gasoline or oil? Show or explain how you know your answer is right. Millie will run out of oil first bec because in 8 days she uses both pints of oil, but in 10 days she uses both 5 gallon continers.

Student F

Millie's Business

This problem gives you the chance to

- *use measures of gallons, pints, and days to solve a mathematical problem*
- *work accurately with numbers (including fractions)*
- *explain the math in your answer*

Millie has a lawn-mowing business. She uses $\frac{1}{4}$ pint of oil each day and $1\frac{1}{2}$ gallons of gasoline each day. She has these supplies on her shelf:

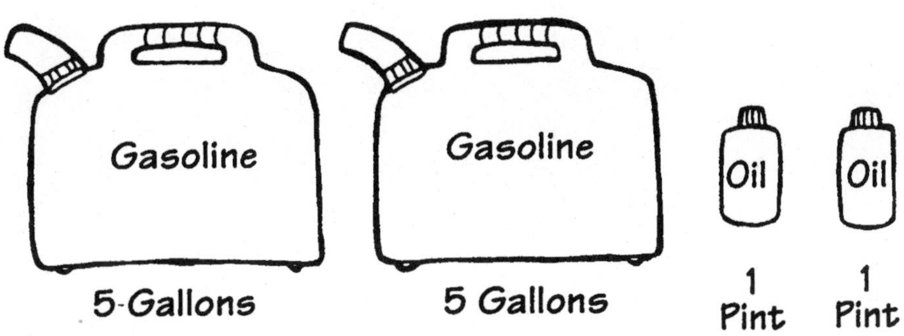

Which will she run out of first, gasoline or oil? Show or explain how you know your answer is right. The gasoline would run out first. In 6 days you waist 9 gallons of gasoline and in 8 days you waist 2 pints of oil.

Task 8

Overview

List all possible combinations.

Solve a problem within given constraints.

Seating Groups

Short Task

Task Description

This task asks students to find all the ways two boys and two girls can sit at four desks, given the constraints that a girl must sit next to a boy and a girl must sit opposite a boy. The important mathematical concept students work with in the task is combinatorics, or listing all possible combinations.

Assumed Mathematical Background

Students should have had some experience making systematic lists of all possible combinations.

Core Elements of Performance

- list all possible combinations
- solve a problem within given constraints

Circumstances

Grouping:	Students work to complete an individual written response.
Materials:	No special materials are needed for this task.
Estimated time:	15 minutes

Name Date

Seating Groups

This problem gives you the chance to

- *show all the ways two boys and two girls can sit together at four tables if they have to obey certain rules for how they can arrange themselves*

Students sit in groups of four in Mrs. Cruz's class. To the right is part of a classroom map that shows Group 1.

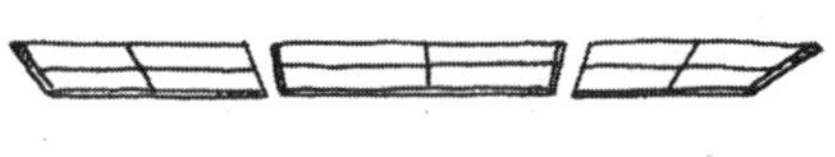

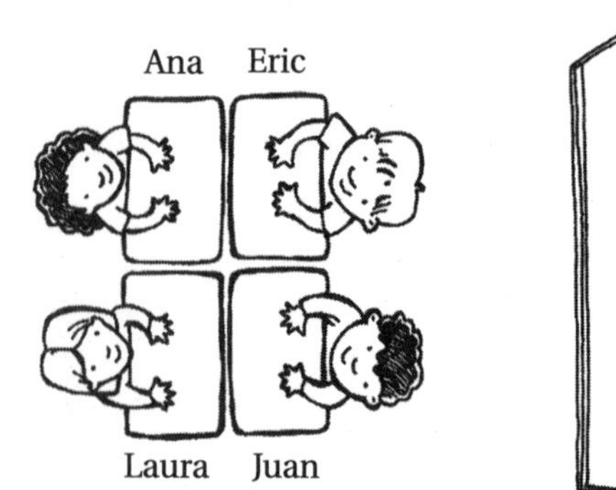

The students in the class agree that they want to spread the boys and girls out evenly. The class made these rules:

1. Each group of four should have two boys and two girls.

2. If you are a girl, the person next to you should be a boy and the person straight across from you should be a boy.

3. If you are a boy, the person next to you should be a girl and the person straight across from you should be a girl.

Two girls, Laura and Ana, and two boys, Juan and Eric, want to be in Group 1. On a sheet of paper, show all the possible ways they can sit as a group. Remember to follow the rules created by the class.

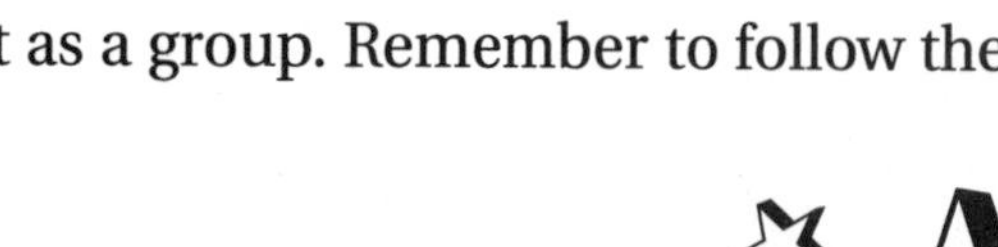

© The Regents of the University of California

A Sample Solution

Task **8**

Laura Juan

Eric Ana

Ana Juan

Eric Laura

Juan Laura

Ana Eric

Eric Laura

Ana Juan

Laura Eric

Juan Ana

Ana Eric

Juan Laura

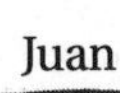

Juan Ana

Laura Eric

Eric Ana

Laura Juan

Task

Using this Task

This task uses the context of classroom seating groups, one with which most elementary school students will have some familiarity, to assess a student's ability to list all possible combinations. While the context is a familiar school setting, the task is a challenging and nonroutine one in which students need to think about a number of constraints as they attempt to solve the problem.

Read the task aloud to students. Discuss what it means to say that if you are a boy, the person next to you has to be a girl, and so on.

Extensions

You may wish to extend this task by using some of the ideas that follow.

- How many ways could two boys and two girls sit in a group of four if there were no rules governing the arrangements?
- Three girls, Laura, Ana, and Misti, and three boys, Eric, Juan, and Gabriel, want to play a game. They need to make teams that have three players on a team. They agreed that each team would have at least one boy and at least one girl on it. What are all the ways they could arrange themselves into teams?

Characterizing Performance

Task

This section offers a characterization of student responses and provides indications of the ways in which the students were successful or unsuccessful in engaging with and completing the task. The descriptions are keyed to the *Core Elements of Performance.* Our global descriptions of student work range from "The student needs significant instruction" to "The student's work meets the essential demands of the task." Samples of student work that exemplify these descriptions of performance are included below, accompanied by commentary on central aspects of each student's response. These sample responses are *representative;* they may not mirror the global description of performance in all respects, being weaker in some and stronger in others.

The characterization of student responses for this task is based on these *Core Elements of Performance:*

1. List all possible combinations.
2. Solve a problem within given constraints.

Descriptions of Student Work

The student needs significant instruction.

These papers show, at most, an attempt to show some seating arrangements for the four students. Typically, they do not consistently adhere to the constraints of boys and girls sitting opposite each other and next to each other. They either show only a few correct arrangements, or they show a combination of correct and incorrect arrangements. It is clear from the work that little or no attention was given to the constraints.

Student A

This response shows an attempt at finding seating arrangements. The response shows four of the eight correct arrangements, but it also shows three arrangements that do not fit the constraints of the task.

Student B

This response shows four arrangements that ignore the constraints of the problem, three correct arrangements, and a repeat of one of the correct arrangements.

Task

The student needs some instruction.

These papers provide evidence of an organized approach to finding possible seating arrangements. They will have at least four arrangements that fit within the constraints of the problem and not more than one incorrect arrangement.

Student C

This response shows five of the possible arrangements and one arrangement that ignores the constraints of the problem.

Student D

This response shows four of the possible arrangements. There are no incorrect arrangements or repeats, but only half of the possible arrangements are shown.

The student's work needs to be revised.

These responses contain no incorrect arrangements. They show most of the possible arrangements and may include a repetition of an arrangement.

Student E

This response shows six of the possible arrangements with no incorrect arrangements or repeats.

Student F

This response shows seven of the possible arrangements and it contains one repeated arrangement.

The student's work meets the essential demands of the task.

These responses show all eight correct arrangements and no incorrect arrangements or repeats.

Student G

This response shows all eight possible arrangements with no incorrect arrangements and no repeats.

Seating Groups ■ Student Work

Student A

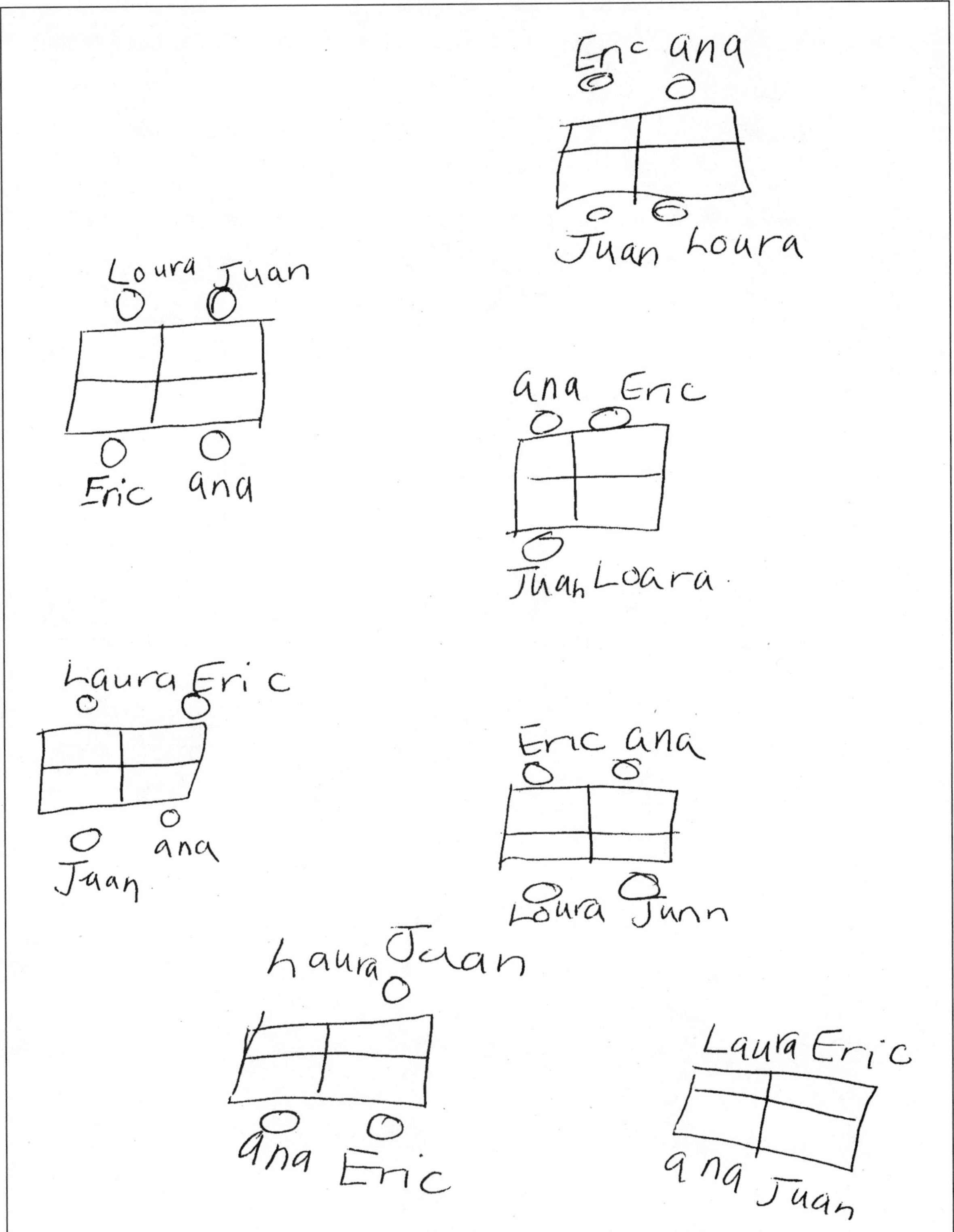

Seating Groups ▪ Student Work

Student B

Eric	Juan
Anna	Laura

Laura	Juan
Anna	Eric

Juan	Anna
Laura	Eric

Laura	Eric
Juan	Anna

Juan	Laura
Eric	Anna

Eric	Laura
Anna	Juan

Juan	Anna
Laura	Eric

Laura	Anna
Juan	Eric

Student C

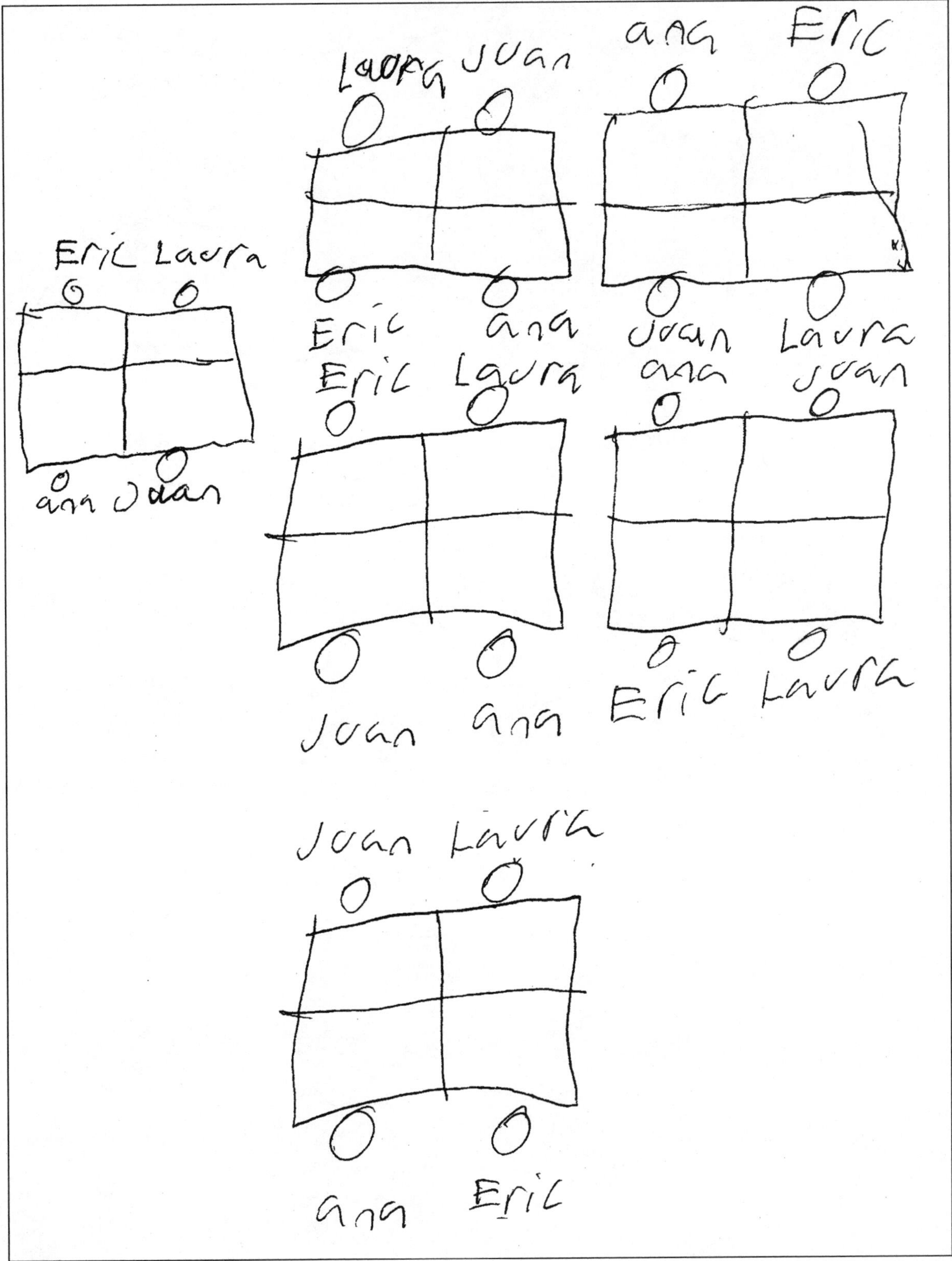
Laura Juan
Eric ana
ana Eric
Juan Laura
Eric Laura
ana Juan
Eric Laura
Juan ana
ana Juan
Eric Laura
Juan Laura
ana Eric

Student D

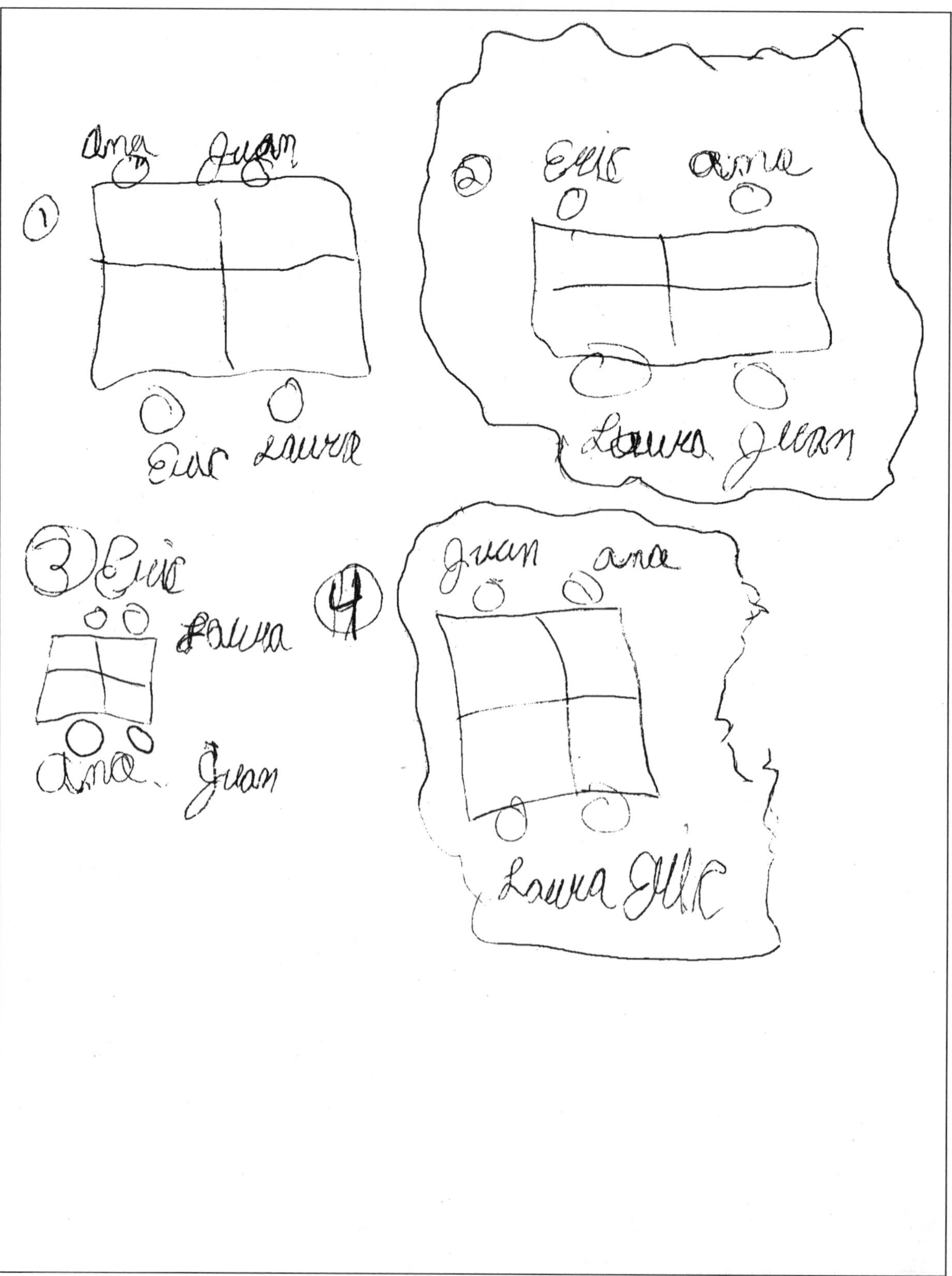

Student E

1.

Laura	Juan
Eric	Ana

2.

Eric	Ana
Laura	Juan

3

Laura	Eric
Juan	Anna

4.

Juan	Ana
Laura	Eric

5

Eric	Laura
Ana	Juan

6.

Ana	Juan
Eric	Laura

Seating Groups ■ Student Work

Student F

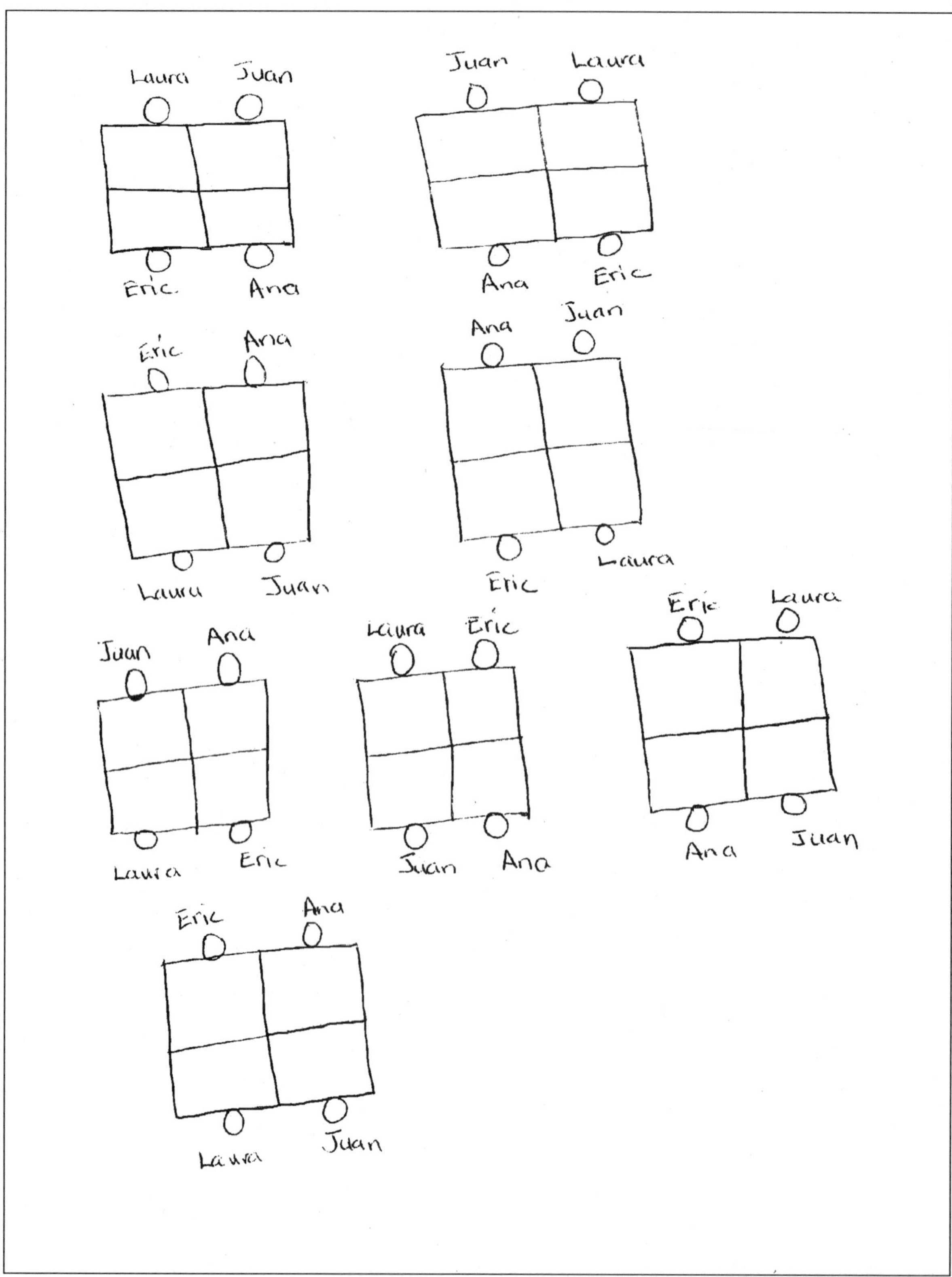

Student G

Laura Eric
Juon Ana
Ana Eric
Juen Laura
LuaraJuan
Erie Ana
Ana Juan
Erie Lugra
Eric Ana
Lugra Juan
Juan And
Luara Enic
JuanLuara
Ana Eric
Eric Lugra
Ana Juan

Task 9

Overview

Interpret data in a bar graph.

Make comparisons based on the data.

Fourth Graders

Short Task

Task Description

This task gives students a double-bar graph showing the number of boys and girls in three fourth-grade classrooms. It asks students to write three or more comparisons based on the data in the graph.

Assumed Mathematical Background

The task assumes that students have had prior experience reading graphs and interpreting data.

Core Elements of Performance

- interpret data in a double-bar graph
- write statements making comparisons based on the data

Circumstances

Grouping:	Students work to complete an individual written response.
Materials:	No special materials are needed for this task.
Estimated time:	15 minutes

Name Date

Fourth Graders

This problem gives you the chance to

- *interpret data in the bar graph*
- *write three comparisons based on that data*

This graph shows how many boys and girls are in each of the three fourth-grade classes at Everest School.

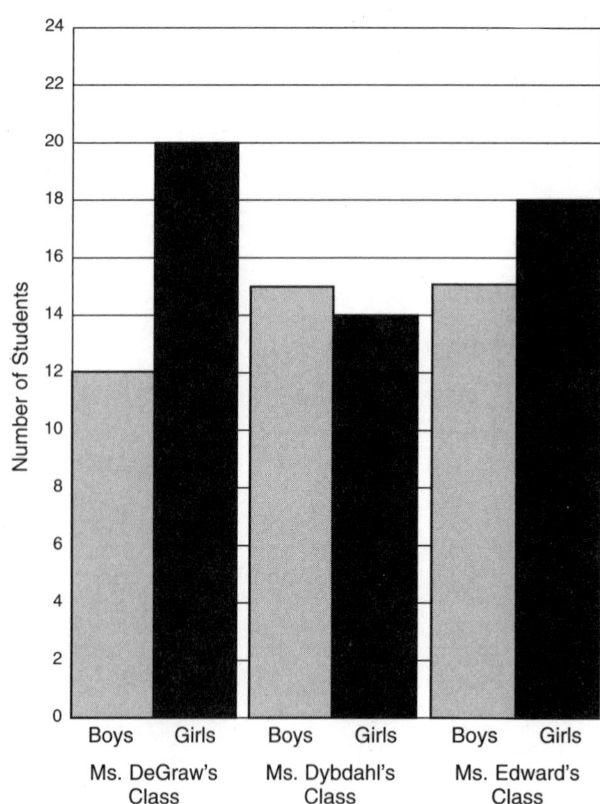

On a sheet of paper, write at least three true comparisons based on this graph. You may compare students within a class or you may compare classes.

© The Regents of the University of California

A Sample Solution

Task

There are many comparisons that can be made. Here are some of them.

- Ms. DeGraw's class has eight more girls than boys (20 girls and only 12 boys).
- Ms. DeGraw's class has more girls than either of the other classes.
- Ms. Dybdahl and Ms. Edward both have the same number of boys in their classes.
- Of the three classes, Ms. Edward's class is the biggest.
- Ms. DeGraw's class has both the most girls and the fewest boys of any of the classes.
- In the three classes combined, there are more girls than boys (52 girls and 42 boys).

Any statement that makes accurate comparisons is acceptable.

Using This Task

This task asks students to interpret data from a graph and use it to make comparisons. Read the task aloud to students. Check to be sure they understand what comparisons are. You may ask them to make comparisons of two things in the classroom to be sure that the concept of comparing is understood.

Extensions

You may wish to extend this task by using some of the ideas that follow.

- Bring in graphs from newspapers or other sources and ask students to interpret them and to make comparisons based on the data.
- Have students conduct a survey of two or more classes, graph the results from each survey, and then compare the graphs. It is interesting to do surveys in a kindergarten class and a fifth-grade class and then compare the younger and older students.

Characterizing Performance

Task

This section offers a characterization of student responses and provides indications of the ways in which the students were successful or unsuccessful in engaging with and completing the task. The descriptions are keyed to the *Core Elements of Performance.* Our global descriptions of student work range from "The student needs significant instruction" to "The student's work meets the essential demands of the task." Samples of student work that exemplify these descriptions of performance are included below, accompanied by commentary on central aspects of each student's response. These sample responses are *representative;* they may not mirror the global description of performance in all respects, being weaker in some and stronger in others.

The characterization of student responses for this task is based on these *Core Elements of Performance:*

1. Interpret data in a double-bar graph.
2. Write statements making comparisons based on the data.

Descriptions of Student Work

The student needs significant instruction.

These responses make some attempt to state information contained in the graph. They do not make comparisons, or they may make inaccurate comparisons, or there may be only one comparison.

Student A

This response tells how many boys and how many girls are in each of the three classes. It contains two inaccurate statements, that there are $14\frac{1}{2}$ boys in Ms. Edward's and Ms. Dybdahl's classes, and that the total (total of what is unclear) is 29. There are no comparisons.

Student B

This response shows one accurate comparison, but there are no other statements.

Task

The student needs some instruction.

These papers contain three statements about the graph, at least one of which is a comparison about the data in the graph. The other two statements may simply state data found in the graph or they may be vague. All three statements are accurate.

Student C

This response shows three statements. One of them comments on the numbers going up the vertical axis of the graph and the other two are vague comparisons about the data.

The student's work needs to be revised.

These responses make three statements based on data from the graph. At least two of the statements are accurate comparisons. The third statement is either an accurate statement that is not a comparison, or it is a comparison that contains a minor error.

Student D

This response shows two accurate comparisons. The third comparison says, "Ms. Edwards class hases the most boys in all 3 4th grades." This is not entirely accurate, as Ms. Edward and Ms. Dybdahl both have the same number of boys in their classes. They do, however, have more boys in their classes than Ms. DeGraw has in her class.

The student's work meets the essential demands of the task.

These responses make three accurate comparisons based on data from the graph.

Student E

This response shows three accurate comparisons.

Student A

I now That thers 12 Boys
oh, Ms Degraws class thers
20 girls. on the class on
Ms DyBDahL's thers 14 girls
and 14 and ½ boys on ms EDwar
DS' class thers 18 girls and
14 ½ boys so the total is
29.

Student B

There are 6 more girls in Ms. Degraw's class than Ms. Dybdahl's does in her class.

Student C

The first class has <u>way</u> more girls than boys The bar numbers go by twos, not like most graphs. In all most all the classes girls have more of a population.

Student D

Ms. Degarws class hases the most girls in the all 3 4th grades, Ms. Edwards class hases the most boys in all 3 4th grades. Ms. Dybdahl's class hases the Least a mout of Girls in a 4th grades

Student E

(1) In Ms Degraws class there are more girls, tha boys.

(2) The boys in Ms Dybdahl's class have the same amount in Ms Edwards class.

(3) In Ms Degraws class they have more girls than hoys but in Ms Dybdahl's class they have more boys than girls.

Task 10

Overview

Visualize a shape rotated 90 degrees.
Draw the rotated shape.

Rotating Shapes

Short Task

Task Description

In this task students are asked to look at a given shape and to visualize what that shape would look like if it were rotated 90 degrees. Then they draw the rotated shape.

Assumed Mathematical Background

Students should have had some experience with rotational symmetry.

Core Elements of Performance

- construct an accurate rotation of an 8-sided polygon

Circumstances

Grouping:	Students work to complete an individual written response.
Materials:	No special materials are needed for this task.
Estimated time:	5 minutes

Name Date

Rotating Shapes

This problem gives you the chance to

- *draw an accurate rotation of a geometric shape*

Zoila rotated Figure A 90° to the right around the pin at the center of the grid.

Figure A

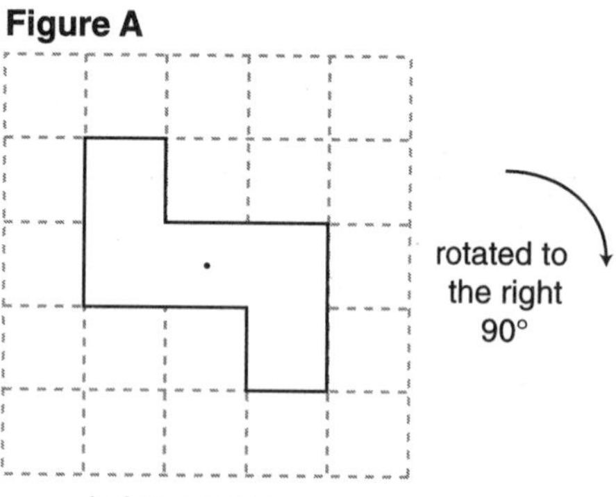

before rotating . . .

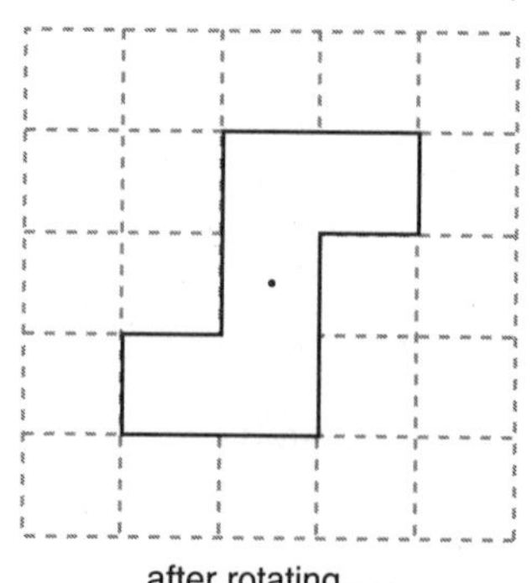

after rotating . . .

Look at Figure B. Rotate this shape 90° to the right around the pin. Draw the rotated shape on the grid.

Figure B

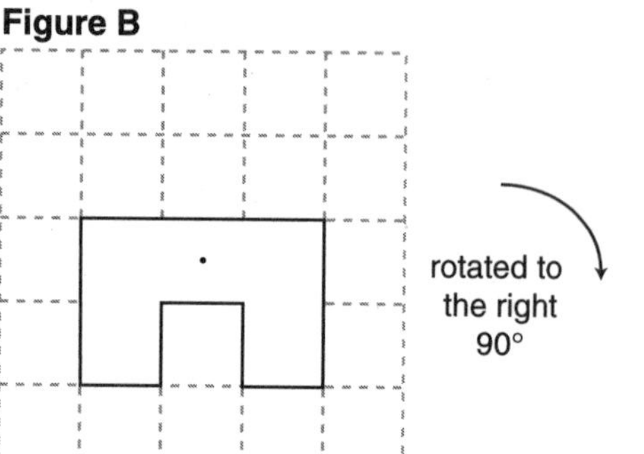

before rotating . . .

after rotating . . .

© The Regents of the University of California

A Sample Solution

Figure B

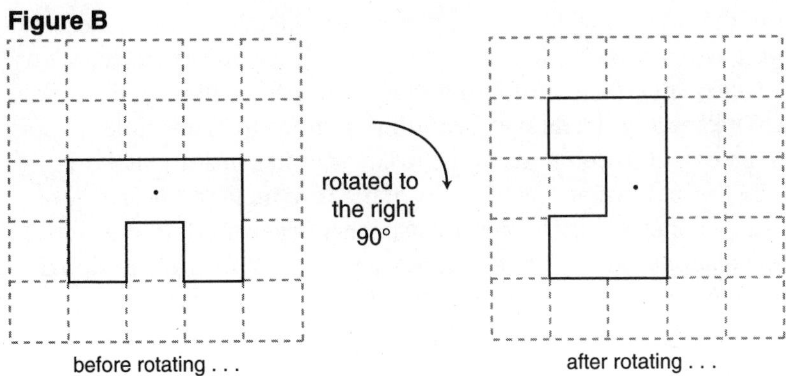

Task

Characterizing Performance

This section offers a characterization of student responses and provides indications of the ways in which the students were successful or unsuccessful in engaging with and completing the task. The descriptions are keyed to the *Core Elements of Performance.* Our global descriptions of student work range from "The student needs significant instruction" to "The student's work meets the essential demands of the task." Samples of student work that exemplify these descriptions of performance are included below, accompanied by commentary on central aspects of each student's response. These sample responses are *representative;* they may not mirror the global description of performance in all respects, being weaker in some and stronger in others.

The characterization of student responses for this task is based on this *Core Element of Performance:*

1. Construct an accurate rotation of an 8-sided polygon.

Descriptions of Student Work

The student needs significant instruction

These papers show, at most, evidence of understanding that the task is to rotate the geometric shape.

Typically, these papers will draw the shape inaccurately, or will rotate the shape 180°, or not rotate it at all.

Student A

This response shows that the shape has been rotated 180°.

The student needs some instruction.

These papers provide evidence of understanding how to rotate the geometric shape 90°. These responses may show that the shape was rotated 90° to the left, rather than to the right. These responses may or may not show that the shape is also moved to the left or right, or it is moved up or down on the grid.

Student B

This response shows that the shape has been rotated 90° to the left, rather than to the right. The shape has also been moved to the left on the grid.

The student's work needs to be revised.

These papers show a rotation of the shape 90° to the right. The shape is moved to the left or right, or it is moved up or down on the grid.

Student C

This response shows that the shape has been rotated 90° to the right. The shape has been moved to the right on the grid.

The student's work meets the essential demands of the task.

These papers show the correct shape rotated 90° to the right on the grid, with the shape's center in the same location on the grid as it was in the original drawing.

Student D

This response shows that the shape has been rotated 90° to the right. The center of the shape is located in the same place on the grid as it was in the original drawing.

Student A

Rotating Shapes

This problem gives you the chance to

- *draw an accurate rotation of a geometric shape*

Zoila rotated Figure A 90° to the right around the pin at the center of the grid.

Figure A

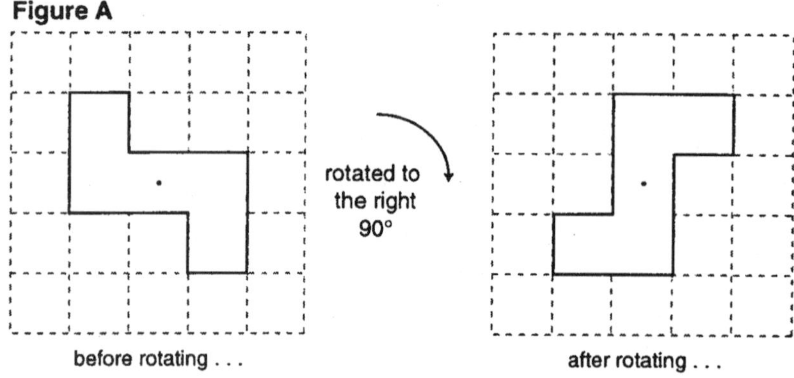

Look at Figure B. Rotate this shape 90° to the right around the pin. Draw the rotated shape on the grid.

Figure B

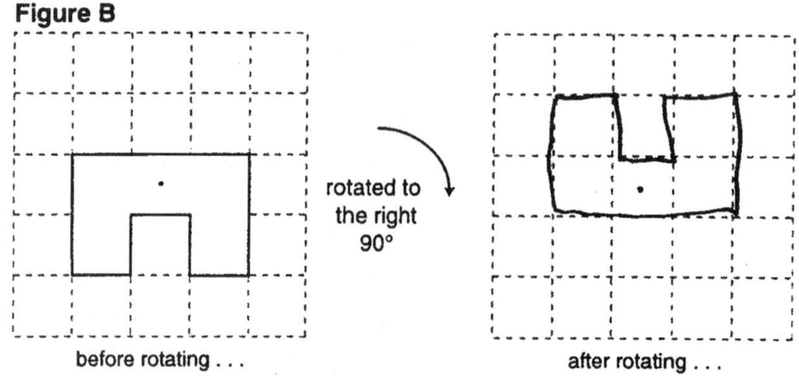

Student B

Rotating Shapes

This problem gives you the chance to

- *draw an accurate rotation of a geometric shape*

Zoila rotated Figure A 90° to the right around the pin at the center of the grid.

Figure A

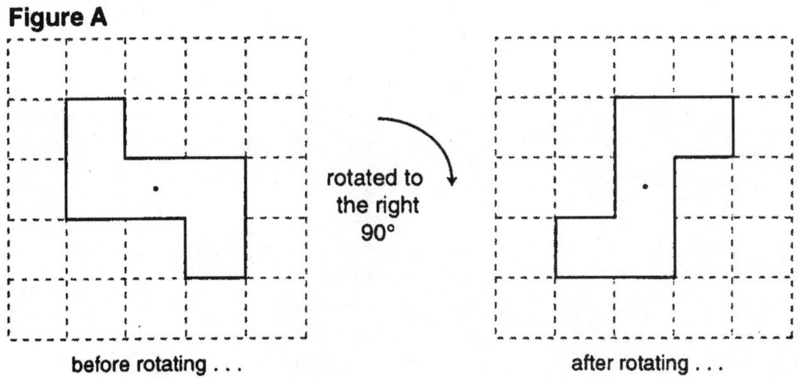

Look at Figure B. Rotate this shape 90° to the right around the pin. Draw the rotated shape on the grid.

Figure B

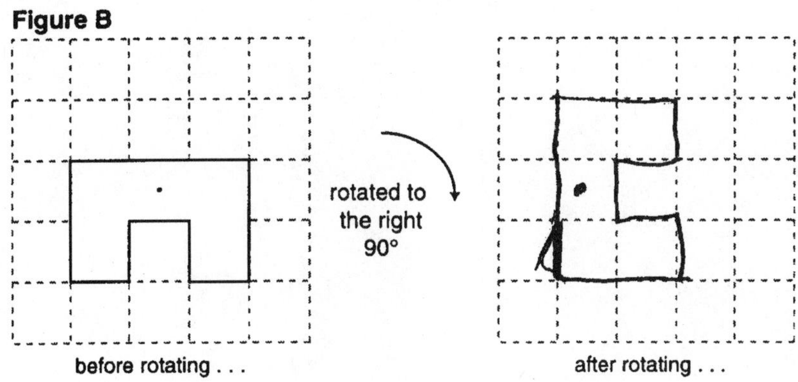

Student C

Rotating Shapes

This problem gives you the chance to

- *draw an accurate rotation of a geometric shape*

Zoila rotated Figure A 90° to the right around the pin at the center of the grid.

Figure A

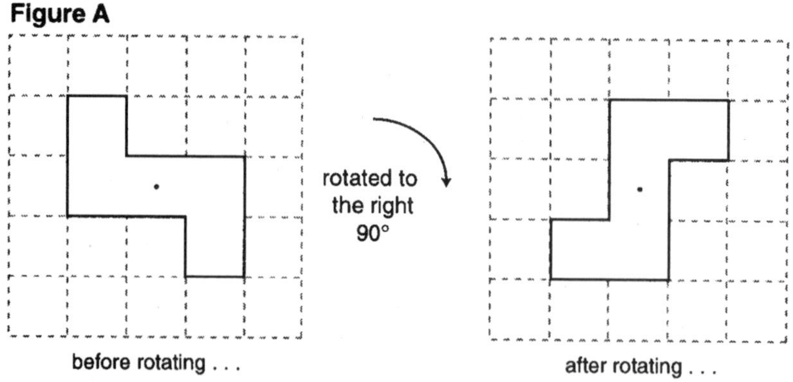

Look at Figure B. Rotate this shape 90° to the right around the pin. Draw the rotated shape on the grid.

Figure B

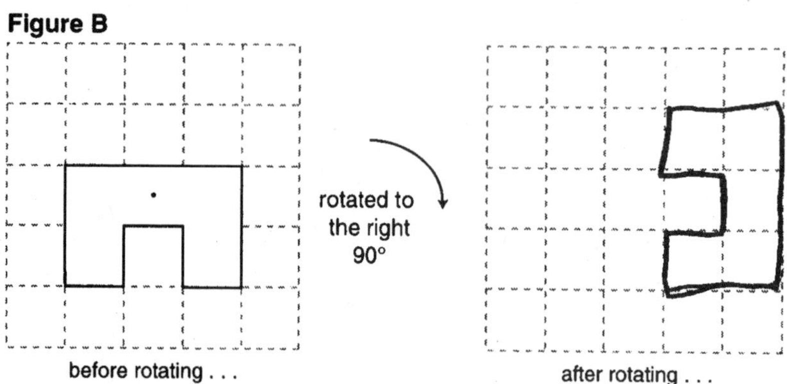

Student D

Rotating Shapes

This problem gives you the chance to

- *draw an accurate rotation of a geometric shape*

Zoila rotated Figure A 90° to the right around the pin at the center of the grid.

Figure A

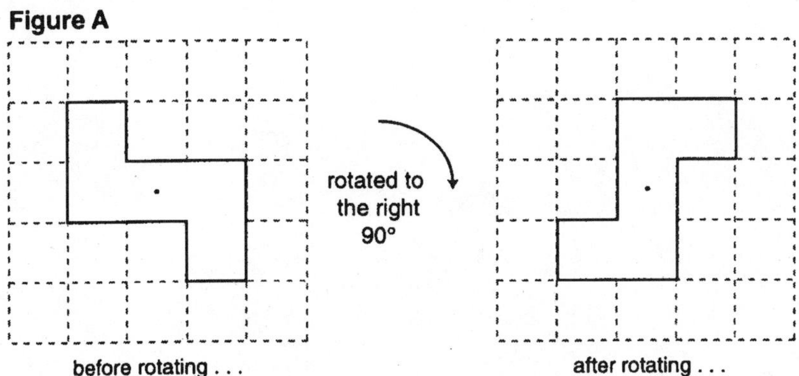

Look at Figure B. Rotate this shape 90° to the right around the pin. Draw the rotated shape on the grid.

Figure B

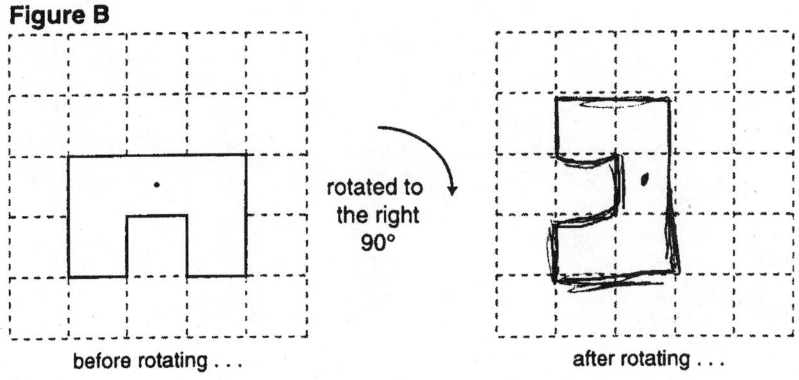

Task 11

Overview

Interpret information on a map.
Use a map scale.

Going to Grandma's

Short Task

Task Description

In this task students use a map scale and information in the prompt to find a location on a map.

Assumed Mathematical Background

Students should have had some experience reading and interpreting information from a map. They should have some prior experience with map scale.

Core Elements of Performance

- interpret information on a map
- use a map scale

Circumstances

Grouping:	Students work to complete an individual written response.
Materials:	No special materials are needed for this task.
Estimated time:	5 minutes

Name Date

Going to Grandma's

This problem gives you the chance to

- *interpret information on a map*

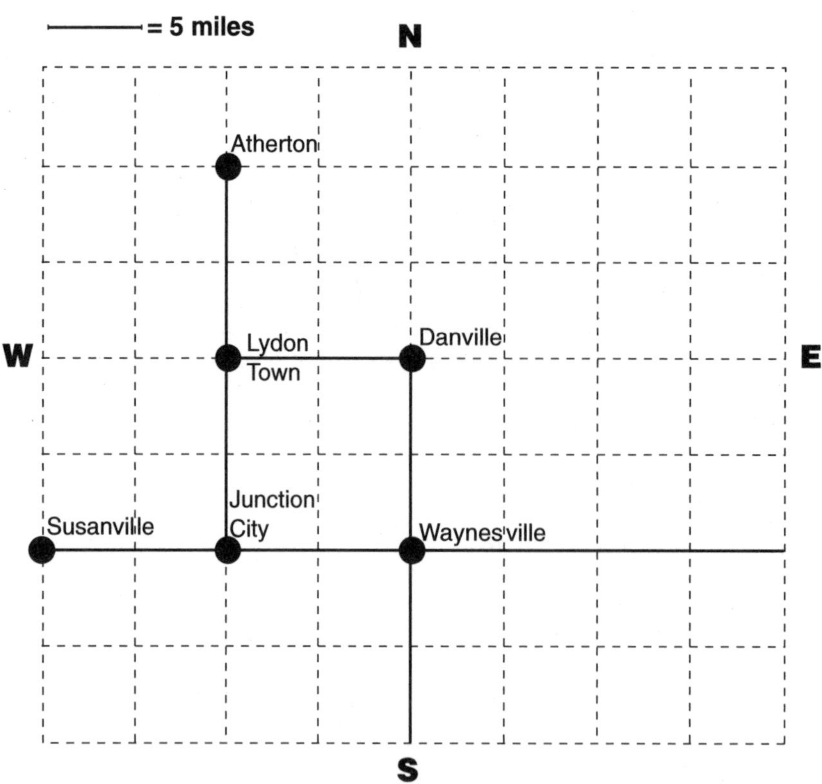

When the Hermans stopped for lunch in Junction City, Emma's dad told her, "Only ten miles north and we'll be at grandma's."

1. Look at the map. Where does Emma's grandma live? ______________

____

Then Emma's dad said, "Maybe we should stop by your Uncle Joe's house first. He lives in Susanville."

2. Look at the map. How far and in which direction does Uncle Joe live from Junction City? ______________

© The Regents of the University of California

A Sample Solution

Task

1. Grandma lives in Lydon Town. Any other answer is incorrect.
2. Uncle Joe lives in Susanville. Susanville is 10 miles west of Junction City.

Task

Characterizing Performance

This section offers a characterization of student responses and provides indications of the ways in which the students were successful or unsuccessful in engaging with and completing the task. The descriptions are keyed to the *Core Elements of Performance.* Our global descriptions of student work range from "The student needs significant instruction" to "The student's work meets the essential demands of the task." Samples of student work that exemplify these descriptions of performance are included below, accompanied by commentary on central aspects of each student's response. These sample responses are *representative;* they may not mirror the global description of performance in all respects, being weaker in some and stronger in others.

The characterization of student responses for this task is based on these *Core Elements of Performance:*

1. Interpret information on a map.
2. Use a map scale.

Descriptions of Student Work

The student needs significant instruction.

These papers name any town other than Lydon Town for question 1. They may also measure the distance inaccurately or give the wrong direction for question 2.

Student A

While Waynesville is the correct distance from Junction City, it is 10 miles east, rather than 10 miles north of Junction City. There is no response to question 2.

This student needs some instruction.

These responses show an attempt to interpret the information on the map. The answer includes an accurate response to question 1. The response to question 2, however, is often unsuccessful.

Task

Student B

This student correctly answers question 1. The student's answer to question 2 is "20 miles." This is not only the incorrect distance, but it is also incomplete because the direction to Susanville is not mentioned.

The student's work needs to be revised.

Responses at this level make a reasonable interpretation of the map based on the map scale. The response, however, may contain an error pertaining to the direction of Susanville.

Student C

This student answers question 1 appropriately, but makes an error in question 2. The student mistakenly writes down "east" rather than west as the direction to Susanville.

The student's work meets the essential demands of the task.

These papers indicate that Grandma lives in Lydon Town and Uncle Joe's house is 10 miles west of Junction City.

Student D

This response correctly answers the questions.

Student A

This problem gives you the chance to

- *interpret information on a map*

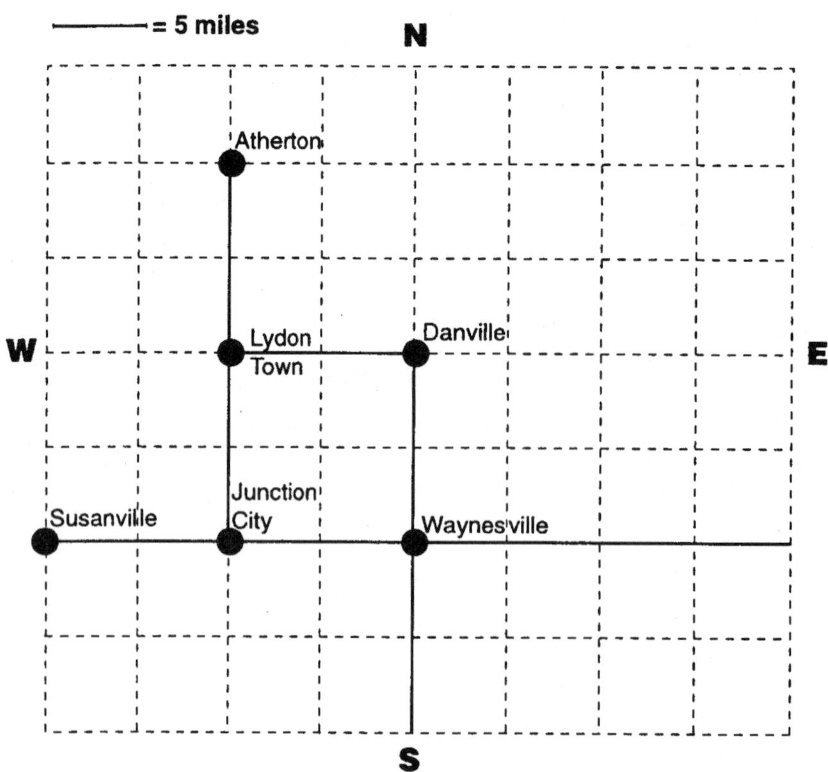

When the Hermans stopped for lunch in Junction City, Emma's dad told her, "Only ten miles north and we'll be at grandma's."

1. Look at the map. Where does Emma's grandma live? Waynesville

Then Emma's dad said, "Maybe we should stop by your Uncle Joe's house first. He lives in Susanville."

2. Look at the map. How far and in which direction does Uncle live from Junction City? ____________

Student B

This problem gives you the chance to

- *interpret information on a map*

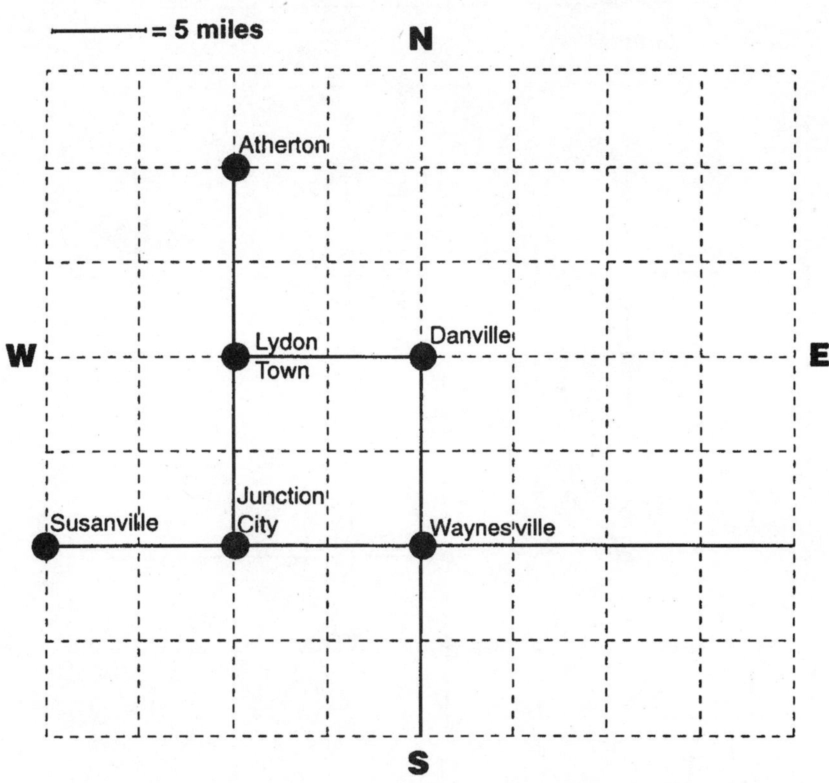

When the Hermans stopped for lunch in Junction City, Emma's dad told her, "Only ten miles north and we'll be at grandma's."

1. Look at the map. Where does Emma's grandma live? Lydon Town

Then Emma's dad said, "Maybe we should stop by your Uncle Joe's house first. He lives in Susanville."

2. Look at the map. How far and in which direction does Uncle Joe live from Junction City? 20 Miles

Student C

This problem gives you the chance to

- *interpret information on a map*

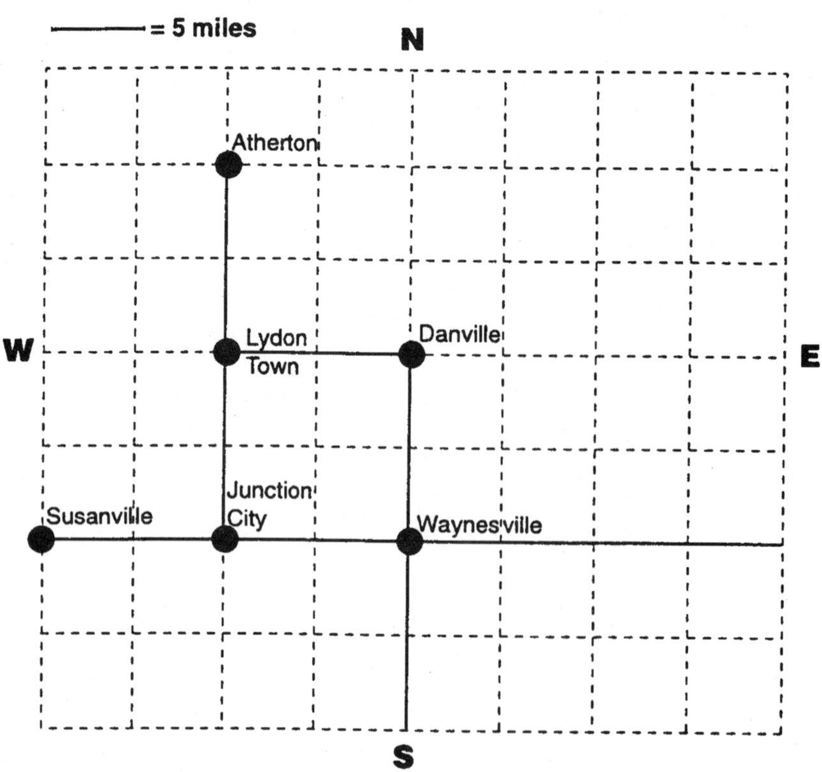

When the Hermans stopped for lunch in Junction City, Emma's dad told her, "Only ten miles north and we'll be at grandma's."

1. Look at the map. Where does Emma's grandma live? Lydon Town

Then Emma's dad said, "Maybe we should stop by your Uncle Joe's house first. He lives in Susanville."

2. Look at the map. How far and in which direction does Uncle Joe live from Junction City? 10 miles - East

Student D

This problem gives you the chance to

- *interpret information on a map*

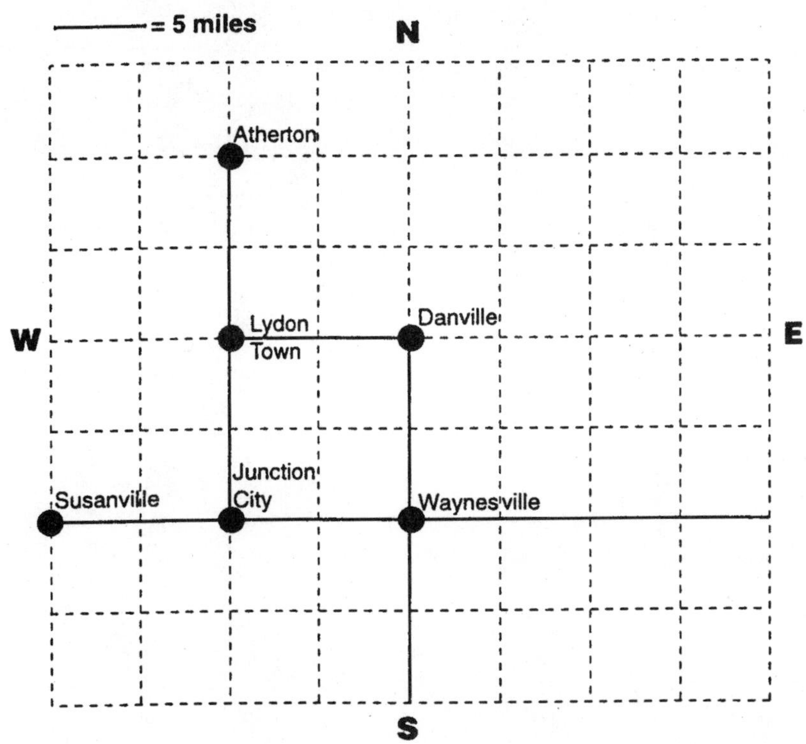

When the Hermans stopped for lunch in Junction City, Emma's dad told her, "Only ten miles north and we'll be at grandma's."

1. Look at the map. Where does Emma's grandma live? Lydon Town

Then Emma's dad said, "Maybe we should stop by your Uncle Joe's house first. He lives in Susanville."

2. Look at the map. How far and in which direction does Uncle Joe live from Junction City? Go 10 miles West to get to Uncle Joe's house.

Task 12

Drawing a Spinner

Overview

Interpret data in a table.

Analyze results of spinning a spinner.

Predict what the spinner might look like.

Short Task

Task Description

In this task students use information from a table showing results of spinning a spinner. They are asked to predict what the spinner might look like.

Assumed Mathematical Background

Students should have had some experience with probability experiments.

Core Elements of Performance

- analyze and interpret data in a table
- draw reasonable conclusions about probability based on available data

Circumstances

Grouping:	Students work to complete an individual written response.
Materials:	No special materials are needed for this task.
Estimated time:	15 minutes

Name Date

Drawing a Spinner

This problem gives you the chance to

- *interpret data in a table*
- *show what the spinner that was used to get that data could look like*
- *explain your thinking*

Connie experimented with a spinner. She did 3 experiments. Each time she spun her spinner 20 times. She made this table to show her results.

	Red	Yellow	Green
First 20 Spins	𝍸 II	𝍸 𝍸	III
Second 20 Spins	III	𝍸 𝍸 II	𝍸
Third 20 Spins	𝍸	𝍸 IIII	𝍸 I

1. Use the circle below to draw a picture of what the colors on her spinner probably looked like.
2. Explain why you think this is what her spinner looked like. ______________________

____

____

____

____

____

© The Regents of the University of California

A Sample Solution

Task 12

1.

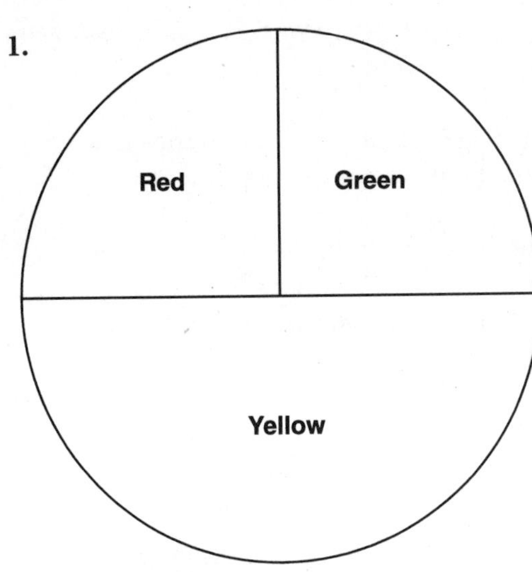

2. The spinner Connie used probably looked like this. Students could create this by counting how many times the spinner landed on each color. It landed on yellow 31 times out of 60, and 31 is very close to half of 60. So the yellow could be about half of Connie's spinner. It landed on red 15 times and on green 14 times, and 14 and 15 are almost the same. If students look at Connie's table they would see that red came up more than green on the first 20 spins, but green came up more than red on the middle 20 and on the last 20 spins. This information is enough to justify that green and red should have about the same amount of space on the spinner.

Using this Task

Make sure that students understand that they are to work individually on this task. Distribute copies of the task to students.

Ensure that students understand the basic situation, and that they'll be using information in the table Connie made to complete this task.

This task takes about 15 minutes and may be combined, if desired, with other short tasks, for example other 5-, 10-, or 15-minute Balanced Assessment tasks. If this task is combined with other tasks, it is recommended that the entire set of tasks not take more than about 45 minutes.

Extensions

You may wish to extend this task by using some of the ideas that follow.

- Give students tables of results from other spinners and have them draw those spinners.
- Give students pictures of two or three different spinners and a list of statements about them. Students identify which spinner most likely fits each statement. For example:

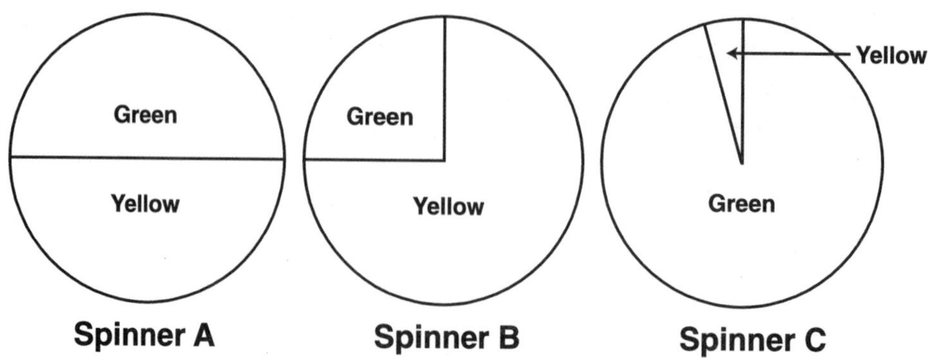

Which spinner is equally likely to land on green and yellow?
Which spinner is most likely to land on yellow?
Which spinner is unlikely to land on yellow?

Characterizing Performance

Task

This section offers a characterization of student responses and provides indications of the ways in which the students were successful or unsuccessful in engaging with and completing the task. The descriptions are keyed to the *Core Elements of Performance.* Our global descriptions of student work range from "The student needs significant instruction" to "The student's work meets the essential demands of the task." Samples of student work that exemplify these descriptions of performance are included below, accompanied by commentary on central aspects of each student's response. These sample responses are *representative;* they may not mirror the global description of performance in all respects, being weaker in some and stronger in others.

The characterization of student responses for this task is based on these *Core Elements of Performance:*

1. Analyze and interpret data in a table.
2. Draw reasonable conclusions about probability based on available data.

Descriptions of Student Work

The student needs significant instruction.

These responses show an attempt, however unsuccessful, to engage with the mathematics of the task. The response may show a spinner of some sort, but it may not pay attention to the data in the table, basing the new spinner on other considerations.

Student A

This response shows an attempt to engage in the task, which is unsuccessful. The spinner is based on an interpretation that does not account for the information in the table.

The student needs some instruction.

These responses provide evidence of an attempt to interpret the data in the table. The interpretation may be unreasonable, incomplete, or the drawing of the spinner may not be consistent with the written explanation.

Task

Student B

This response shows an attempt to interpret the data in the table. The number of times the spinner landed on any given color is miscalculated. (Student interpreted five groups as tens for the yellow portion.) The response states that yellow "is the biggest number so I made it have the biggest space." However, the response gives red almost as much space on the spinner as yellow and yellow has less than half the total space on the spinner.

Student C

This response pays attention to the first section of the table only and bases its spinner drawing on only those twenty spins. The spinner has nearly equal areas for red and yellow and a small space for green because "green only has t(h)ree ta(l)ly marks."

Student D

This response shows an incorrect interpretation of the table. The spinner assigns the most space to red and the explanation says, "its more lik(e)ly for it to land on a big space so it might land on red more."

The student's work needs to be revised.

These responses make a reasonable interpretation of the data in the table. The colors on the spinner may not be apportioned correctly, or the explanation may be weak or incomplete.

Student E

This response shows a spinner that is half yellow and one quarter each red and green. The explanation is incomplete in that it simply says yellow "kept on getting a higher score because it had more space to land on."

Student F

This response shows an attempt to interpret the data in the table that is evidenced by the written explanation, "Yellow got the most so I made that one the biggest." However, the actual drawing of the spinner allocates just slightly over one third to yellow.

The student's work meets the essential demands of the task.

These responses meet the demands of the task. They correctly interpret the table and show a spinner which is $\frac{1}{2}$ yellow, $\frac{1}{4}$ green, and $\frac{1}{4}$ red. The explanation may state that red and green came up about an equal number of times, so they each represent equal parts of the spinner.

Student G

This response shows an accurate interpretation of the table which can be seen in the statement, "Red and green were sort of the same." The spinner shows that the conclusions drawn about probability are reasonable. The spinner is $\frac{4}{8}$ yellow, $\frac{2}{8}$ red, and $\frac{2}{8}$ green.

Student H

This response shows a careful analysis of the data in the table and the spinner shows that reasonable conclusions were drawn as to what the spinner might look like.

Student A

This problem gives you the chance to

- *interpret data in a table*
- *show what the spinner that was used to get that data could look like*
- *explain your thinking*

Connie experimented with a spinner. She did 3 experiments. Each time she spun her spinner 20 times. She made this table to show her results.

	Red	Yellow	Green
First 20 Spins	卌 ||	卌 卌	|||
Second 20 Spins	|||	卌 卌 ||	卌
Third 20 Spins	卌	卌 ||||	卌 |

1. Use the circle below to draw a picture of what the colors on her spinner probably looked like.
2. Explain why you think this is what her spinner looked like. I think my solution is fair because you look a one color and you have to move three spaces before you see the same color.

Student B

This problem gives you the chance to

- *interpret data in a table*
- *show what the spinner that was used to get that data could look like*
- *explain your thinking*

Connie experimented with a spinner. She did 3 experiments. Each time she spun her spinner 20 times. She made this table to show her results.

	Red	Yellow	Green
First 20 Spins	卌 II	卌 卌	III
Second 20 Spins	III	卌 卌 II	卌
Third 20 Spins	卌	卌 IIII	卌 I

1. Use the circle below to draw a picture of what the colors on her spinner probably looked like.
2. Explain why you think this is what her spinner looked like. Yello is the biggest number so I made it have the Biggest space I did the same thing with the other colors except I made a smaller space

Student C

This problem gives you the chance to

- *interpret data in a table*
- *show what the spinner that was used to get that data could look like*
- *explain your thinking*

Connie experimented with a spinner. She did 3 experiments. Each time she spun her spinner 20 times. She made this table to show her results.

	Red	Yellow	Green
First 20 Spins	卌 II	卌 卌	III
Second 20 Spins	III	卌 卌 II	卌
Third 20 Spins	卌	卌 IIII	卌 I

1. Use the circle below to draw a picture of what the colors on her spinner probably looked like.
2. Explain why you think this is what her spinner looked like. Because Yellow and Red has most taly martes and Green only has ~~t~~ tree talymartes

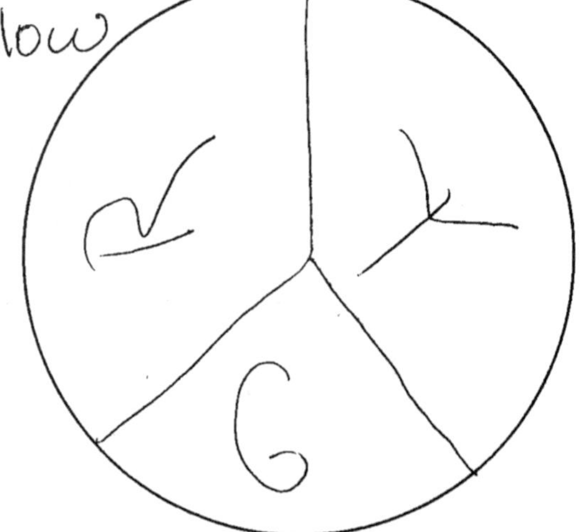

Student D

This problem gives you the chance to

- *interpret data in a table*
- *show what the spinner that was used to get that data could look like*
- *explain your thinking*

Connie experimented with a spinner. She did 3 experiments. Each time she spun her spinner 20 times. She made this table to show her results.

	Red	Yellow	Green
First 20 Spins	𝍸 II	𝍸 𝍸	III
Second 20 Spins	III	𝍸 𝍸 II	𝍸
Third 20 Spins	𝍸	𝍸 IIII	𝍸 I

1. Use the circle below to draw a picture of what the colors on her spinner probably looked like.
2. Explain why you think this is what her spinner looked like. I think its this way because its more likly for it to land on a big space so it might land on Red more.

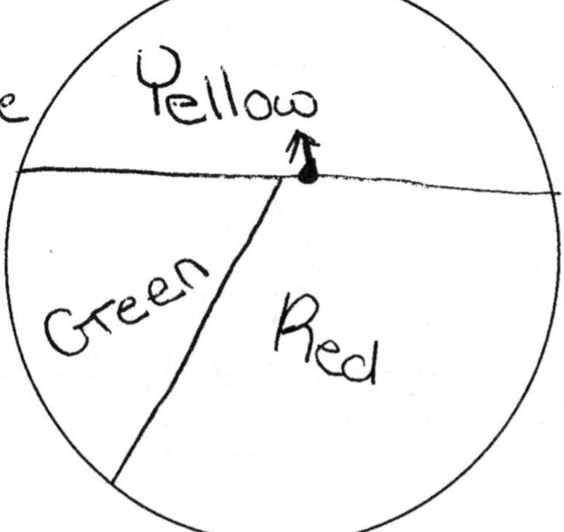

Student E

This problem gives you the chance to

- *interpret data in a table*
- *show what the spinner that was used to get that data could look like*
- *explain your thinking*

Connie experimented with a spinner. She did 3 experiments. Each time she spun her spinner 20 times. She made this table to show her results.

	Red	Yellow	Green
First 20 Spins	𝍸 ||	𝍸 𝍸	|||
Second 20 Spins	|||	𝍸 𝍸 ||	𝍸
Third 20 Spins	𝍸	𝍸 ||||	𝍸 |

1. Use the circle below to draw a picture of what the colors on her spinner probably looked like.
2. Explain why you think this is what her spinner looked like. why yellow kept on getting a higher score was because it had more space to land on.

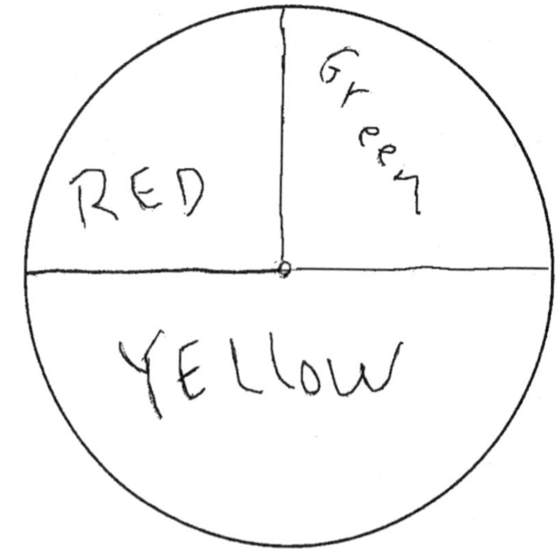

Drawing a Spinner ■ Student Work

Student F

This problem gives you the chance to

- *interpret data in a table*
- *show what the spinner that was used to get that data could look like*
- *explain your thinking*

Connie experimented with a spinner. She did 3 experiments. Each time she spun her spinner 20 times. She made this table to show her results.

	Red	Yellow	Green
First 20 Spins	卌 ||	卌 卌	|||
Second 20 Spins	|||	卌 卌 ||	卌
Third 20 Spins	卌	卌 ||||	卌 |

1. Use the circle below to draw a picture of what the colors on her spinner probably looked like.
2. Explain why you think this is what her spinner looked like. I think it would Look like this because the yellow got the most so I made that one the biggest. The red had the second most I made that the second and the green got the least I Made it even smaller.

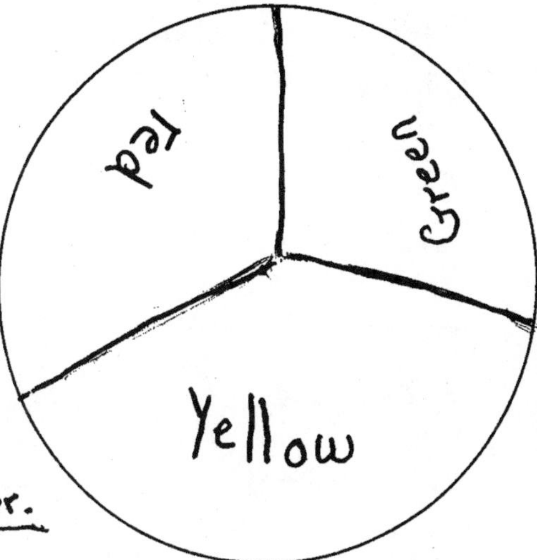

Student G

This problem gives you the chance to

- *interpret data in a table*
- *show what the spinner that was used to get that data could look like*
- *explain your thinking*

Connie experimented with a spinner. She did 3 experiments. Each time she spun her spinner 20 times. She made this table to show her results.

	Red	Yellow	Green
First 20 Spins	~~||||~~ ||	~~||||~~ ~~||||~~	|||
Second 20 Spins	|||	~~||||~~ ~~||||~~ ||	~~||||~~
Third 20 Spins	~~||||~~	~~||||~~ ||||	~~||||~~ |

1. Use the circle below to draw a picture of what the colors on her spinner probably looked like.
2. Explain why you think this is what her spinner looked like. Because Yellow got spun the most and red and green were sort of the same. I did yellow the most because it got more on the graph.

Yellow
Yellow
Red
Red
Yellow
green
green
Yellow

Student H

This problem gives you the chance to

- *interpret data in a table*
- *show what the spinner that was used to get that data could look like*
- *explain your thinking*

Connie experimented with a spinner. She did 3 experiments. Each time she spun her spinner 20 times. She made this table to show her results.

	Red	Yellow	Green
First 20 Spins	𝍸 ‖ 7	𝍸 𝍸 10	||| 3
Second 20 Spins	||| 3	𝍸 𝍸 ‖ 12	𝍸 5
Third 20 Spins	𝍸 5	𝍸 |||| 9	𝍸	6
	15	31	14

1. Use the circle below to draw a picture of what the colors on her spinner probably looked like.
2. Explain why you think this is what her spinner looked like. I think Yellow should be half because it had the most spins on it and Red and Green should be even because there ~~was~~ had only one ~~to~~ spin different.

Yellow Red Green

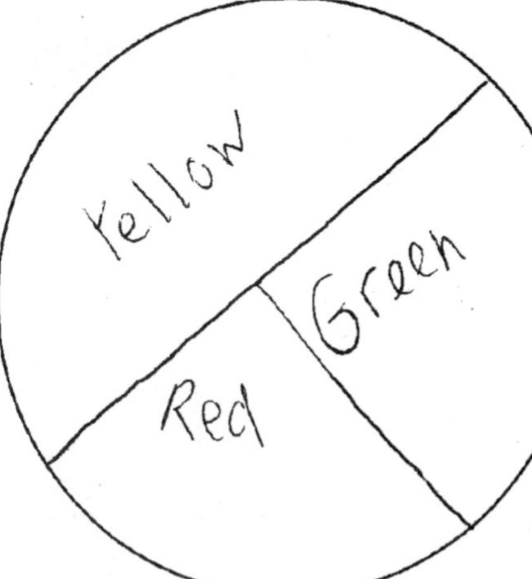

Task 13

Addition Rings

Overview

Demonstrate arithmetic computation skills using addition and subtraction.

Recognize the net change of zero.

Short Task

Task Description

The intent of this problem is to have students demonstrate arithmetic computation skills, specifically the relationship between addition and its inverse, subtraction. We want students to recognize that in this ring, the order in which the addition and subtraction computations are carried out doesn't matter.

Assumed Mathematical Background

The task assumes elementary arithmetic instruction.

Core Elements of Performance

- arrive at a correct result for all computations, regardless of the initial number chosen and its position in the ring
- exhibit an understanding of the arithmetic generalization described by the ring, and the inverse relationship of the arithmetic processes involved

Circumstances

Grouping:	Students work to complete an individual written response.
Materials:	No special materials are needed for this task.
Estimated time:	20 minutes

Name Date

Addition Rings

This problem gives you the chance to

- *use the number ring to make some computations*
- *describe your results*
- *explain why your results came out the way they did*

This is a special kind of number ring. Any number that goes into the top circle gets changed as it goes around to the other circles.

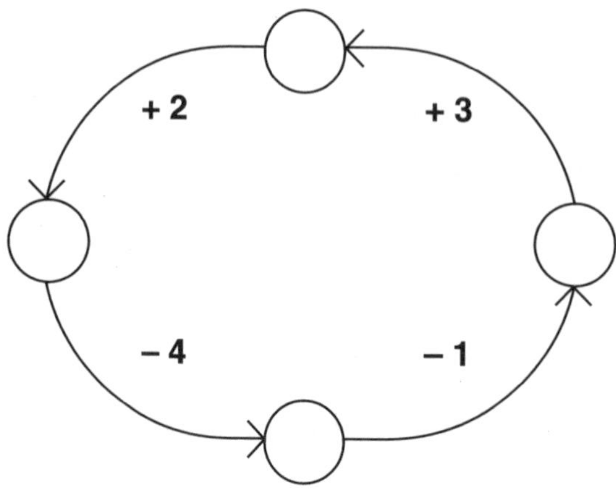

1. Try this ring: Put a number in the top circle. Now move to the left and add 2. Write the answer in the corresponding circle. Continue around the number ring and fill in the circles. What is the result? ______________________________

____

____

© The Regents of the University of California

Name Date

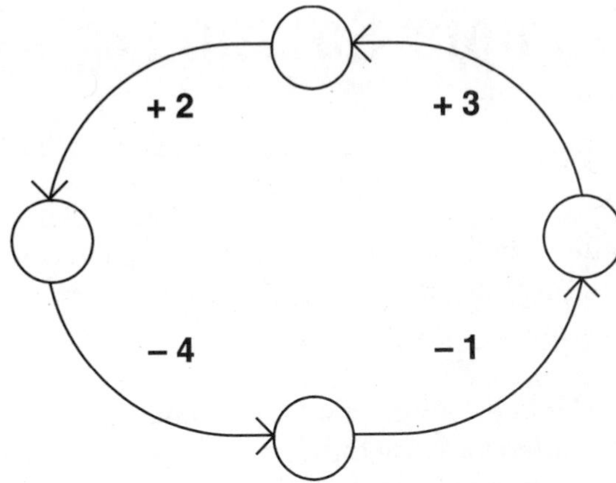

2. Try this again with another number. What is the result?

____

____

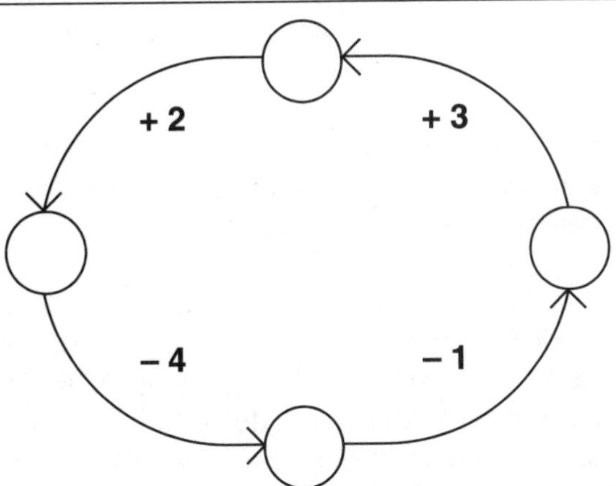

3. Try the ring again, but this time put the number into the bottom circle and follow around the ring. What is the result?

____

____

4. Explain why you think you got these results.

____

____

____

____

© The Regents of the University of California

Task

A Sample Solution

No matter what number is put into the top circle in questions 1 and 2, going through the other three circles and returning to the top produces the same number.

It might be interesting to note that going clockwise through +3 and –1 and going counterclockwise through +2 gives the same result. The second interesting thing to note is that the **final** result does not depend on the direction taken; both result in the net change of 0.

However, in questions 1 and 2, if the original number is 1, 2, or 0, the number at the bottom circle (if moving counterclockwise) turns out to be negative and may cause some anxiety. To avoid this difficulty a larger number should be chosen.

It turns out that starting from the bottom circle in question 3 (with a number greater than 3) does not change the general result: going around the full circle gets you back to the original number (independently of the direction).

These results are all due to the fact that the operations are balanced or counteracted by each other, that is, the combined effect of all the additions is counteracted by the net effect of all the subtractions. In other words, adding 2 and 3 is the same as adding 1 and 4, so adding 2 and 3 and then subtracting 1 and 4 is the same as adding 1 and 4 and subtracting 1 and 4. The operations of adding a number and subtracting the same number have inverse or opposite effect, so the net result is no change.

Using this Task

For informal classroom use, you can adjust the difficulty level in this task by providing more background information to the students. The printed version of the task does not provide students with the basic facts about the Commutative Properties of Addition and Subtraction, nor does it give any hints about what initial number to use.

Extensions

An additional question might be inserted asking students to change direction of movement around the ring; depending on their starting point, however, this may involve them in an initial value that is a negative number.

Task

Characterizing Performance

This section offers a characterization of student responses and provides indications of the ways in which the students were successful or unsuccessful in engaging with and completing the task. The descriptions are keyed to the *Core Elements of Performance.* Our global descriptions of student work range from "The student needs significant instruction" to "The student's work meets the essential demands of the task." Samples of student work that exemplify these descriptions of performance are included below, accompanied by commentary on central aspects of each student's response. These sample responses are *representative;* they may not mirror the global description of performance in all respects, being weaker in some and stronger in others.

The characterization of student responses for this task is based on these *Core Elements of Performance:*

1. Arrive at a correct result for all computations, regardless of the initial number chosen and its position in the ring.
2. Exhibit an understanding of the arithmetic generalization described by the ring, and the inverse relationship of the arithmetic processes involved.

Descriptions of Student Work

The student needs significant instruction.

These responses generally show that an attempt has been made, but they do not arrive at correct answers for any of the questions.

Student A

This response indicates a lack of understanding of the task. The numbers in the rings do not correspond with the numbers given as answers. There seems to be an attempt to add and subtract the same number in order to get the same answer, but this is not explored within the context of the problem.

The student needs some instruction.

These responses generally give evidence of noticing that the number that goes into the ring is the number that comes out, and they may recognize that the result does not depend on the starting point in the ring. There is, however, little or no ability to articulate a reason for the workings of the ring.

Student B

This response clearly indicates that the number the student started with is the same number that he/she ended with. However, the final answer makes it clear that there is no understanding of why this works the way it does.

The student's work needs to be revised.

The response gives evidence of noticing that the number that goes into the ring is the number that comes out, and it may recognize that the result does not depend on the starting point in the ring. However, the explanation of why the ring works like this will be very limited.

Student C

This response clearly demonstrates the working of the ring. The explanation, however, is very limited and needs to be revised to better explain why the ring works like this.

The student's work meets the essential demands of the task.

The response gives evidence of noticing that the number that goes into the ring is the number that comes out, and it may recognize that the result does not depend on the starting point in the ring. The response may be able to adapt if the initial number is too small, causing a negative number at other points in the ring. The description given in question 4 demonstrates the ability to make a rudimentary generalization with respect to the initial position of the number in the ring and/or the inverse relationship of the arithmetic processes involved; all explanations are fully and clearly expressed.

Some responses may mention that it makes no difference whether you go clockwise or counterclockwise around the ring; however, this is not a requirement of a response at this level.

Student D

This response essentially meets the demands of the task. The work on questions 1 through 3 is clear and complete. The explanation given for question 4 clearly explains the inverse relationship of the arithmetic processes by stating "you add five and subtract five."

Student A

Addition Rings

This problem gives you the chance to

- *use the number ring to make some computations*
- *describe your results*
- *explain why your results came out the way they did*

This is a special kind of number ring. Any number that goes into the top circle gets changed as it goes around to the other circles.

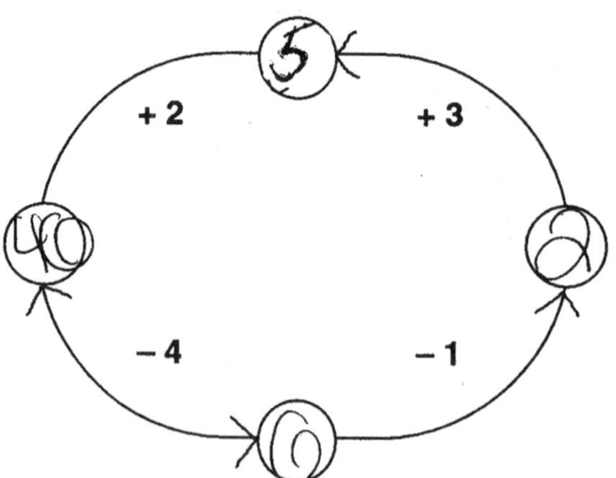

1. Try this ring: Put a number in the top circle. Now move to the left and add 2. Write the answer in the corresponding circle. Continue around the number ring and fill in the circles. What is the result? I put 5+1 – 1 is 5

Student A

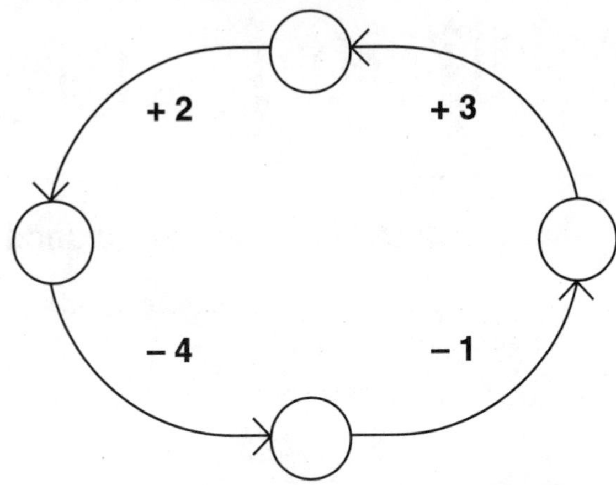

2. Try this again with another number. What is the result?

I Started With 8+0-2.58

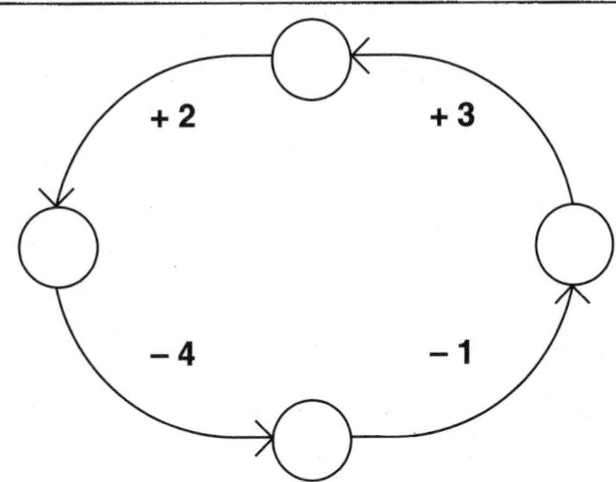

3. Try the ring again, but this time put the number into the bottom circle and follow around the ring. What is the result?

I Put 6+0 - 2=6

4. Explain why you think you got these results.

I, think that it would work because . easy to do.

Student B

Addition Rings

This problem gives you the chance to

- *use the number ring to make some computations*
- *describe your results*
- *explain why your results came out the way they did*

This is a special kind of number ring. Any number that goes into the top circle gets changed as it goes around to the other circles.

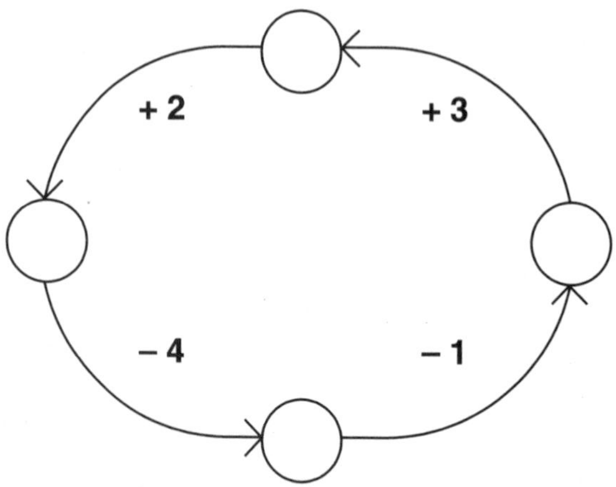

1. Try this ring: Put a number in the top circle. Now move to the left and add 2. Write the answer in the corresponding circle. Continue around the number ring and fill in the circles. What is the result? I Started with 5 and ended withfive perfectly

Student B

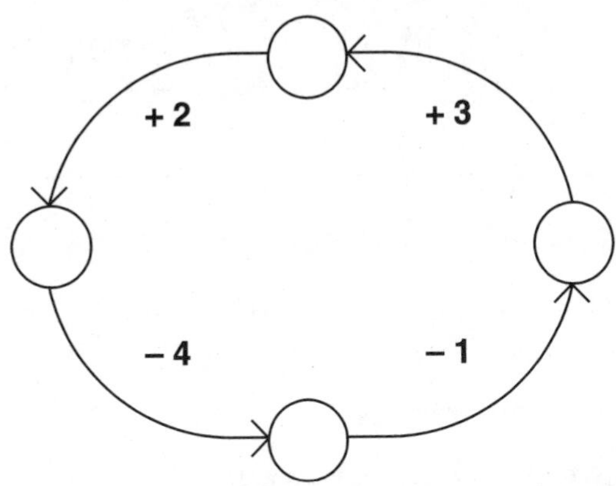

2. Try this again with another number. What is the result?

then I started with 6 and the same thing happend

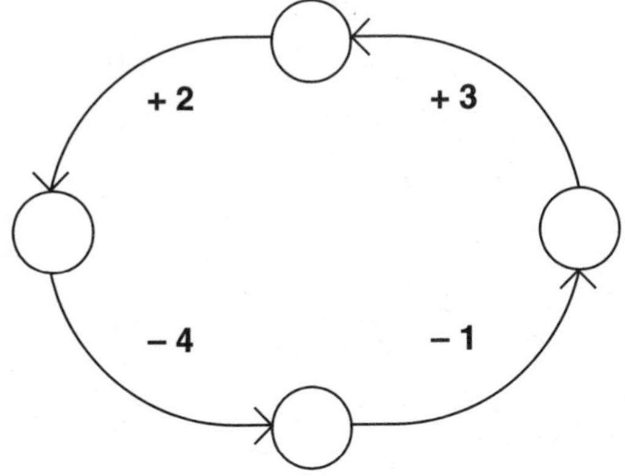

3. Try the ring again, but this time put the number into the bottom circle and follow around the ring. What is the result?

I got it again and I used 7 I got it right too

4. Explain why you think you got these results.

I think this was very fun not too easy and not too hard I like the way you start with one Number and aid with the same.

P.S. I was great

Student C

Addition Rings

This problem gives you the chance to

- *use the number ring to make some computations*
- *describe your results*
- *explain why your results came out the way they did*

This is a special kind of number ring. Any number that goes into the top circle gets changed as it goes around to the other circles.

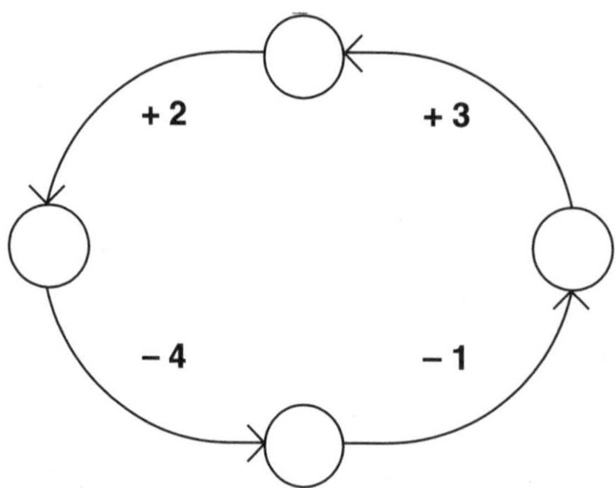

1. Try this ring: Put a number in the top circle. Now move to the left and add 2. Write the answer in the corresponding circle.

   Continue around the number ring and fill in the circles. What is the result? 9 start 9 finsh.

Student C

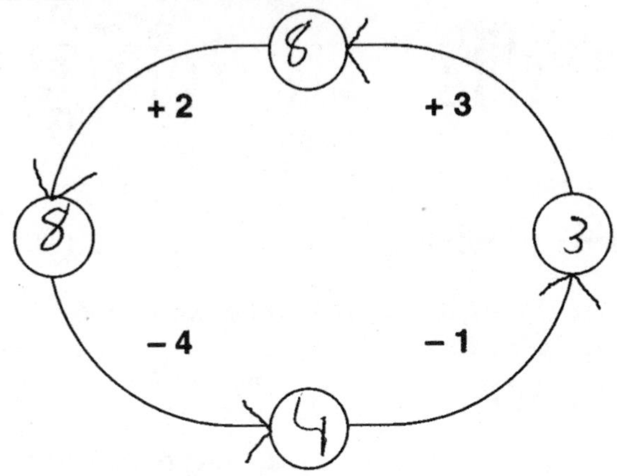

2. Try this again with another number. What is the result?

6 start 6 finish.

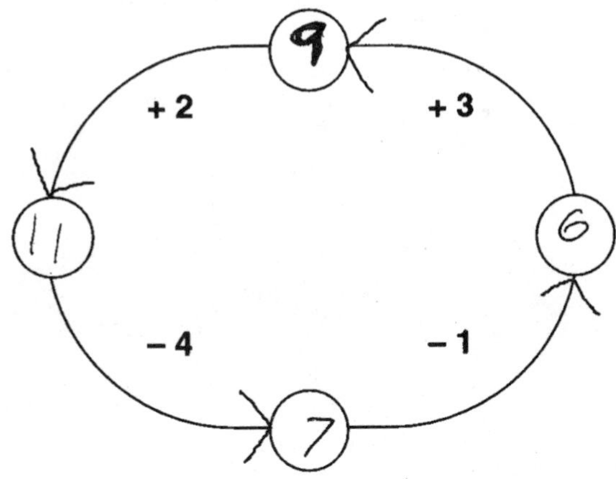

3. Try the ring again, but this time put the number into the bottom circle and follow around the ring. What is the result?

7 start 7 finish.

4. Explain why you think you got these results.

because when you add and saptract you will add with the same number

Student D

Addition Rings

This problem gives you the chance to

- *use the number ring to make some computations*
- *describe your results*
- *explain why your results came out the way they did*

This is a special kind of number ring. Any number that goes into the top circle gets changed as it goes around to the other circles.

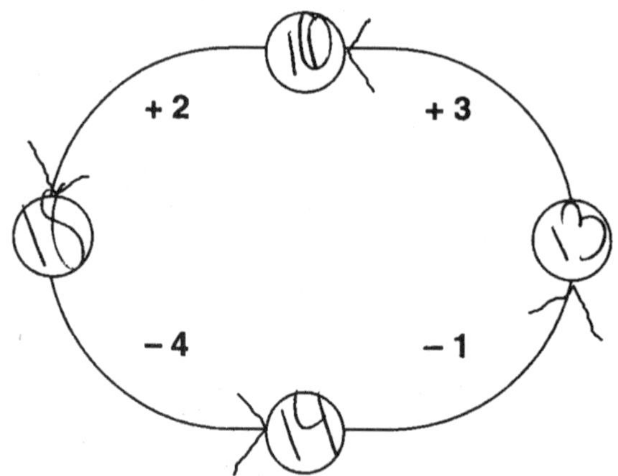

1. Try this ring: Put a number in the top circle. Now move to the left and add 2. Write the answer in the corresponding circle. Continue around the number ring and fill in the circles. What is the result? I started with 16 and got 10

Addition Rings ■ Student Work

Student D

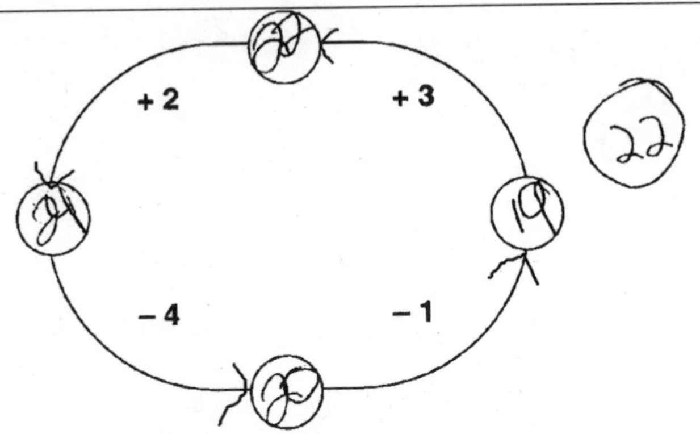

2. Try this again with another number. What is the result?

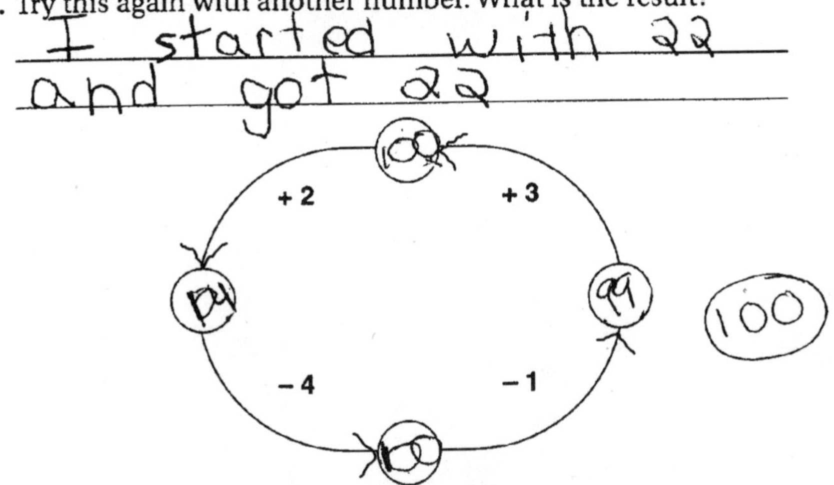

3. Try the ring again, but this time put the number into the bottom circle and follow around the ring. What is the result?

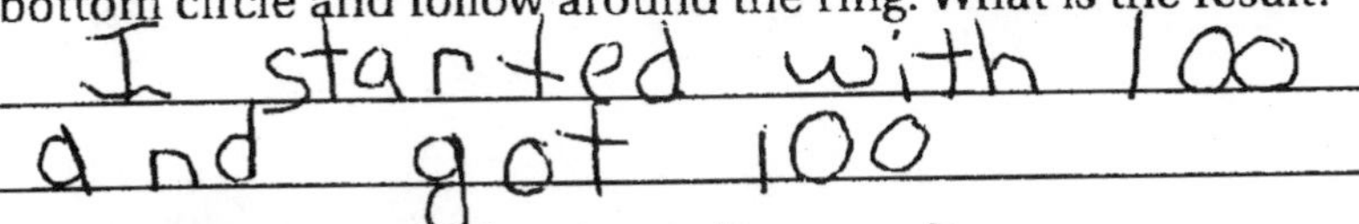

4. Explain why you think you got these results.

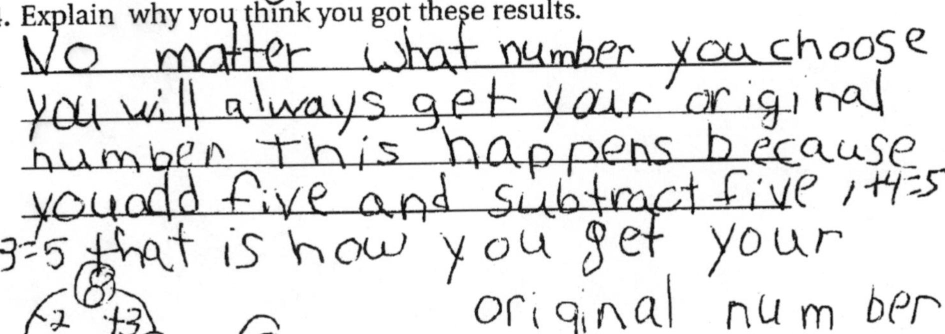

Task 14

Overview

Interpret before-and-after pictures to determine what quantity of a material was consumed.

Anthony Pours Juice

Short Task

Task Description

Students are asked to look at before-and-after pictures of a pitcher of juice and state how much was consumed. Students may approach this problem as one of finding a fractional amount, or they may figure out how many quarts were consumed.

Assumed Mathematical Background

Students should have had some experience applying part-whole reasoning in a variety of concrete situations.

Core Elements of Performance

- decide what portion was consumed by comparing before-and-after pictures

Circumstances

Grouping:	Students work to complete an individual written response.
Materials:	No special materials are needed for this task.
Estimated time:	5 minutes

Name Date

Anthony Pours Juice

This problem gives you the chance to

- *figure out how much juice Anthony poured*
- *explain how you figured it out*

When Anthony took a one-gallon pitcher of juice out of the refrigerator, it looked like this.

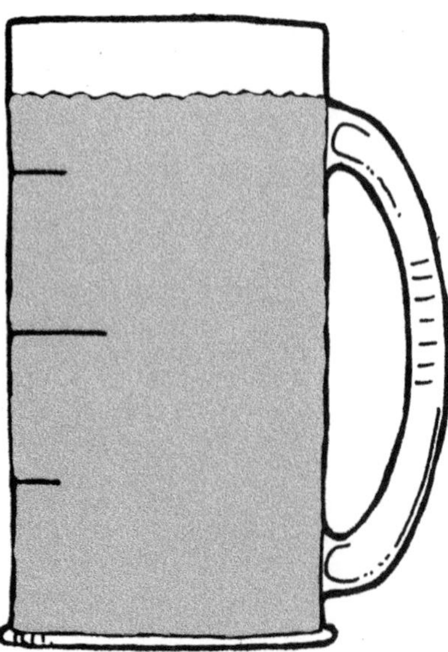

After Anthony poured juice for himself and his friends, the pitcher looked like this.

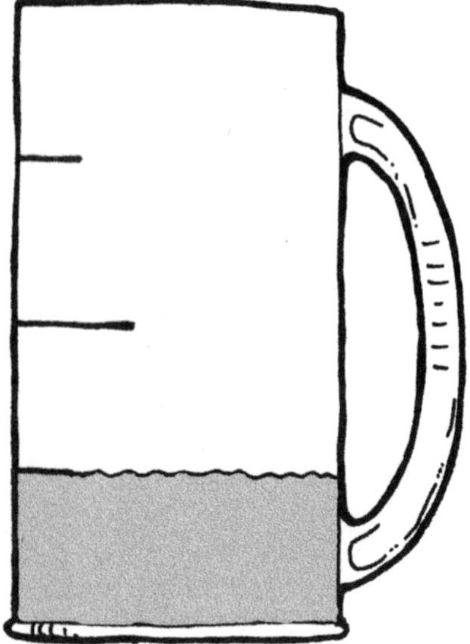

1. How much juice did Anthony pour? ____________________

2. Explain how you figured it out. ____________________

____

____

____

© The Regents of the University of California

A Sample Solution

Task
14

There are two ways in which students can approach this task: they can figure out what proportion of a gallon Anthony poured; or they can figure out how many quarts of juice Anthony poured. Here are sample solutions for both approaches.

Anthony poured $\frac{5}{8}$ of a gallon of juice. Students may figure this out by looking at the pictures. They may notice how much juice there was when Anthony started and how much there was after he poured it. Students could then divide the gallon into eight equal parts. One part was already gone and two parts were left after he poured juice: $\frac{7}{8} - \frac{2}{8} = \frac{5}{8}$. So he poured $\frac{5}{8}$ of a gallon of juice.

OR

If students know that there are four quarts in a gallon, they may also realize that the lines on the pitcher show the quarts. This allows them to see that Anthony poured two and a half quarts of juice because there was already half a quart missing before he poured and there was one quart left when he finished: $4 - 1\frac{1}{2} = 2\frac{1}{2}$ quarts.

Task

Characterizing Performance

This section offers a characterization of student responses and provides indications of the ways in which the students were successful or unsuccessful in engaging with and completing the task. The descriptions are keyed to the *Core Elements of Performance.* Our global descriptions of student work range from "The student needs significant instruction" to "The student's work meets the essential demands of the task." Samples of student work that exemplify these descriptions of performance are included below, accompanied by commentary on central aspects of each student's response. These sample responses are *representative;* they may not mirror the global description of performance in all respects, being weaker in some and stronger in others.

The characterization of student responses for this task is based on this *Core Element of Performance:*

1. Decide what portion was consumed by comparing before-and-after pictures.

Descriptions of Student Work

The student needs significant instruction.

These responses provide evidence of an unsuccessful attempt to figure out the amount of juice Anthony poured.

Student A

This response attempts to engage with the mathematics, but is unsuccessful in finding the amount of juice that Anthony poured.

The student needs some instruction.

These responses are partially successful; however, errors indicating some misunderstanding or confusion are present.

Student B

The response is partially successful; however, it says that Anthony poured "2 and a half of the juice," showing some confusion over what the $2\frac{1}{2}$ represents.

Task

The student's work needs to be revised.

These responses accurately state how much juice Anthony poured, but they do not explain how the students determined this amount.

Student C

This response correctly states that Anthony poured $2\frac{1}{2}$ quarts of juice, but it shows no work and provides no explanation.

The student's work meets the essential demands of the task.

These responses state how much juice Anthony poured and provide a coherent explanation of how they figured it out.

Student D

The paper by Student D is an example of a paper that meets the demands of the task. The response states that Anthony poured $2\frac{1}{2}$ quarts. The explanation is clear and correct.

Student E

This response is also an example of a paper that meets the demands of the task. The response states Anthony poured $\frac{5}{8}$ of a gallon of juice. The explanation is clear and correct.

Student A

Anthony Pours Juice

This problem gives you the chance to

- *figure out how much juice Anthony poured*
- *explain how you figured it out*

When Anthony took a one-gallon pitcher of juice out of the refrigerator, it looked like this.

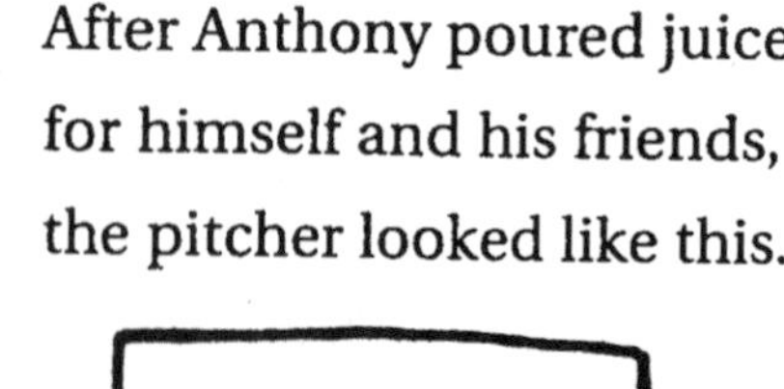

After Anthony poured juice for himself and his friends, the pitcher looked like this.

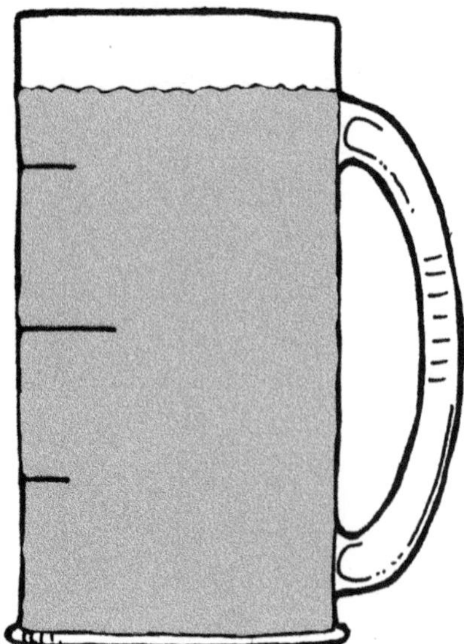

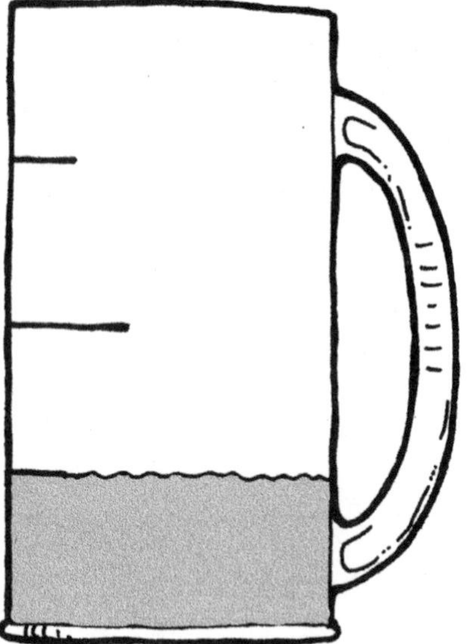

1. How much juice did Anthony pour? Anthony poured $\frac{3}{4}$ and half a quart.

2. Explain how you figured it out. I Figured this out by counting quarts.

Student B

Anthony Pours Juice

This problem gives you the chance to

- *figure out how much juice Anthony poured*
- *explain how you figured it out*

When Anthony took a one-gallon pitcher of juice out of the refrigerator, it looked like this.

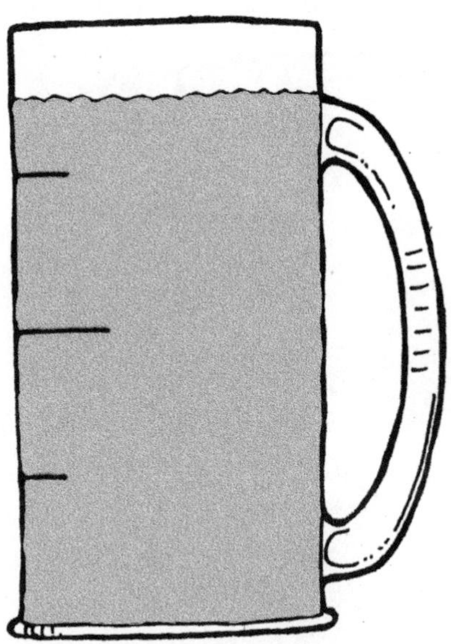

After Anthony poured juice for himself and his friends, the pitcher looked like this.

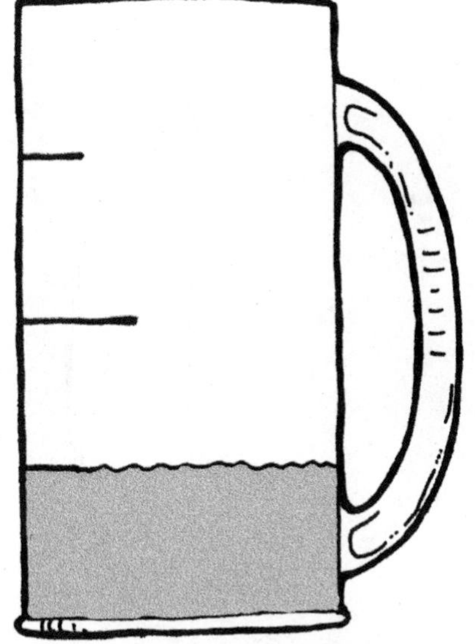

1. How much juice did Anthony pour? ______________

2. Explain how you figured it out. I think Anthony pourd 2 and a half of the Juice! I found this out by looking at the top and And there is one more line left line and going down.

Student C

Anthony Pours Juice

This problem gives you the chance to

- *figure out how much juice Anthony poured*
- *explain how you figured it out*

When Anthony took a one-gallon pitcher of juice out of the refrigerator, it looked like this.

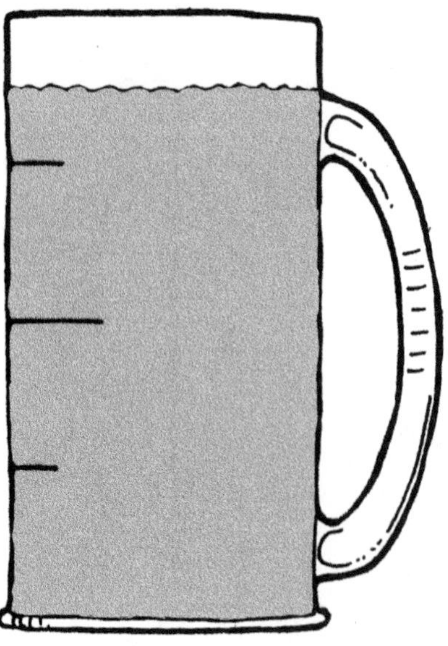

After Anthony poured juice for himself and his friends, the pitcher looked like this.

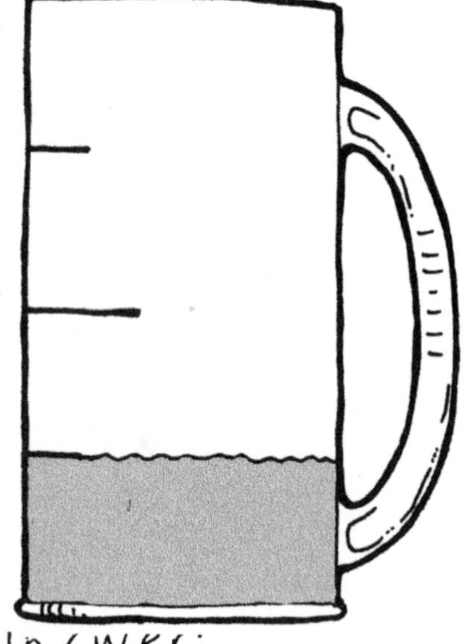

1. How much juice did Anthony pour? answer 2 1/2 Quarts

2. Explain how you figured it out. ______

Student D

Anthony Pours Juice

This problem gives you the chance to

- *figure out how much juice Anthony poured*
- *explain how you figured it out*

When Anthony took a one-gallon pitcher of juice out of the refrigerator, it looked like this.

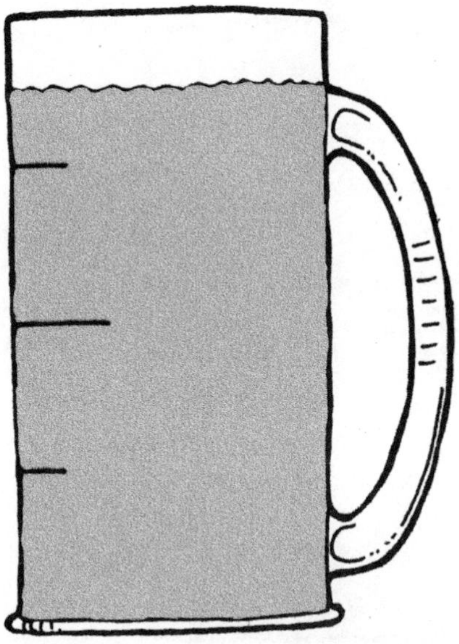

After Anthony poured juice for himself and his friends, the pitcher looked like this.

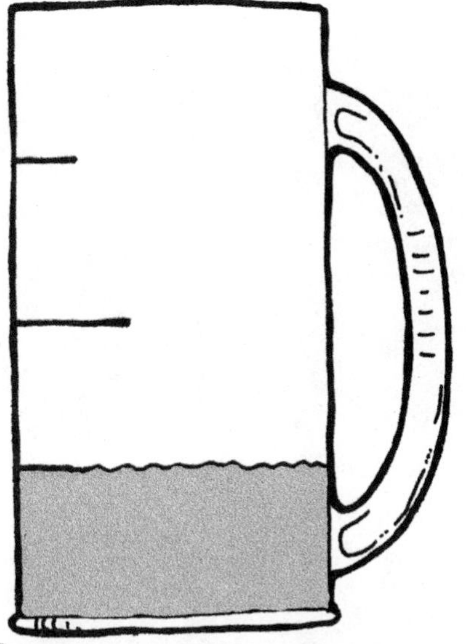

1. How much juice did Anthony pour? answer: 2 quarts and one half of a quart

2. Explain how you figured it out. I got that answer because each mark showed a quart and I just counted the quarts that didn't have juice.

Student E

Anthony Pours Juice

This problem gives you the chance to

- *figure out how much juice Anthony poured*
- *explain how you figured it out*

When Anthony took a one-gallon pitcher of juice out of the refrigerator, it looked like this.

After Anthony poured juice for himself and his friends, the pitcher looked like this.

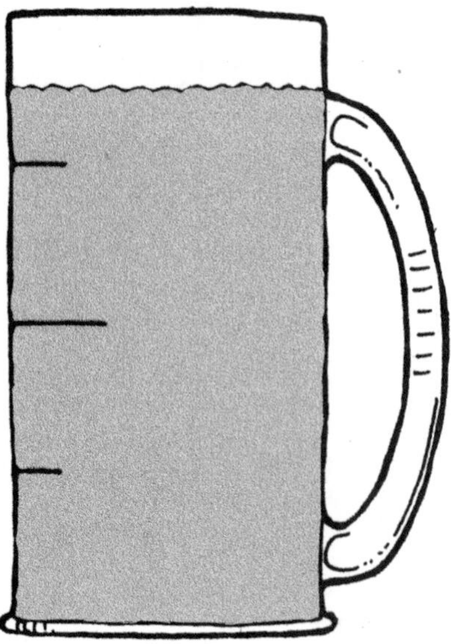

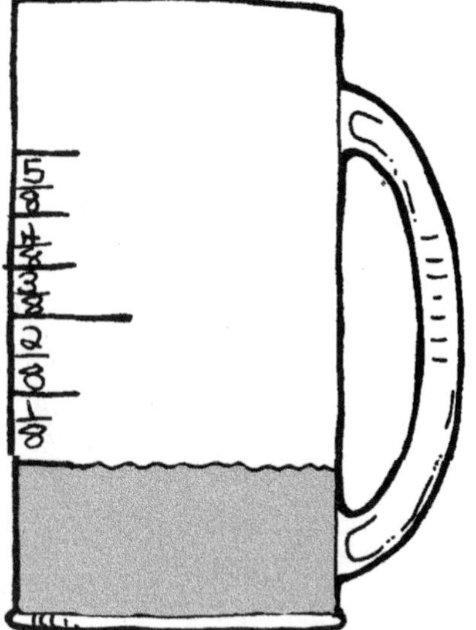

1. How much juice did Anthony pour? $\frac{5}{8}$ of a gallon

2. Explain how you figured it out. Anthony poured $\frac{5}{8}$ of a gallon because there was $\frac{7}{8}$ when he started and only $\frac{2}{8}$ after he poured the juice.

Task 15

Overview

Find a geometric pattern.
Extend the pattern to finish a drawing.

Tile Pattern

Short Task

Task Description

In this task students are asked to look at a partially completed drawing of a tile pattern and to finish drawing the pattern. The task involves students in identifying and extending geometric patterns.

Assumed Mathematical Background

It is assumed that students have had opportunities to look for and extend geometric patterns. Experience with pattern blocks and other geometric manipulatives is helpful.

Core Elements of Performance

- identify a geometric pattern
- extend a geometric pattern

Circumstances

Grouping:	Students work to complete an individual written response.
Materials:	No special materials are needed for this task.
Estimated time:	5 minutes

Name Date

Tile Pattern

This problem gives you the chance to

- *notice and extend a geometric pattern*

Look at this tile pattern. It isn't finished yet. Finish drawing the pattern on the grid. On some parts of the edge, you may be able to show only half a tile.

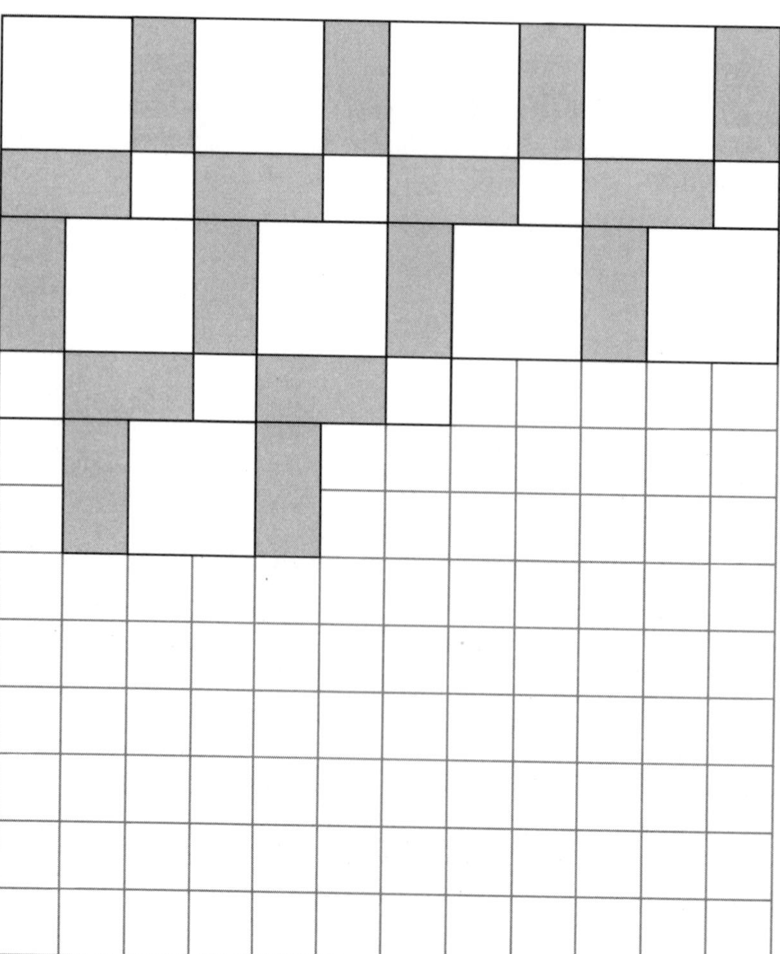

© The Regents of the University of California

A Sample Solution

Task

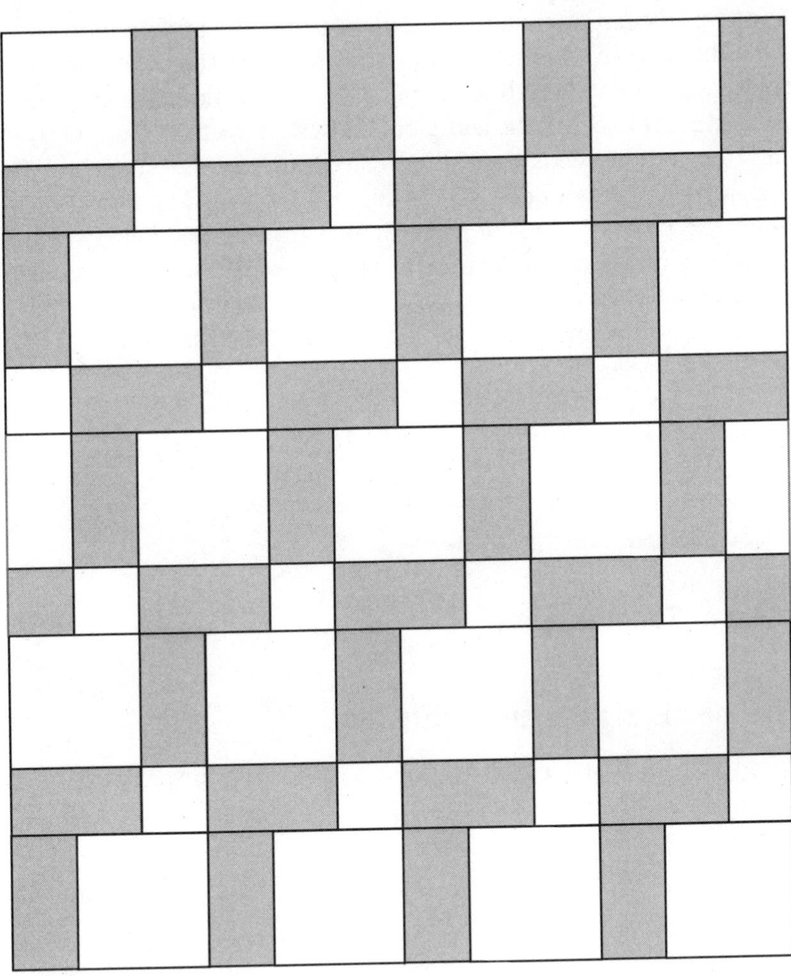

Task

Characterizing Performance

This section offers a characterization of student responses and provides indications of the ways in which the students were successful or unsuccessful in engaging with and completing the task. The descriptions are keyed to the *Core Elements of Performance.* Our global descriptions of student work range from "The student needs significant instruction" to "The student's work meets the essential demands of the task." Samples of student work that exemplify these descriptions of performance are included below, accompanied by commentary on central aspects of each student's response. These sample responses are *representative;* they may not mirror the global description of performance in all respects, being weaker in some and stronger in others.

The characterization of student responses for this task is based on these *Core Elements of Performance:*

1. Identify a geometric pattern.
2. Extend a geometric pattern.

Descriptions of Student Work

The student needs significant instruction.

These papers make an unsuccessful attempt to extend the tile pattern.

Students A and B

Both papers show that an attempt has been made to extend the pattern, but that attempt has not been successful.

The student's work meets the essential demands of the task.

These papers successfully extend the tile pattern. They may contain some minor errors in extending the pattern.

Student C

This response shows that the pattern has been correctly extended to cover the entire grid.

Student D

This response shows that the pattern is extended with one minor error on the right side.

Student A

Tile Pattern

This problem gives you the chance to

- *notice and extend a geometric pattern*

Look at this tile pattern. It isn't finished yet. Finish drawing the pattern on the grid. On some parts of the edge, you may be able to show only half a tile.

Student B

Tile Pattern

This problem gives you the chance to

- *notice and extend a geometric pattern*

Look at this tile pattern. It isn't finished yet. Finish drawing the pattern on the grid. On some parts of the edge, you may be able to show only half a tile.

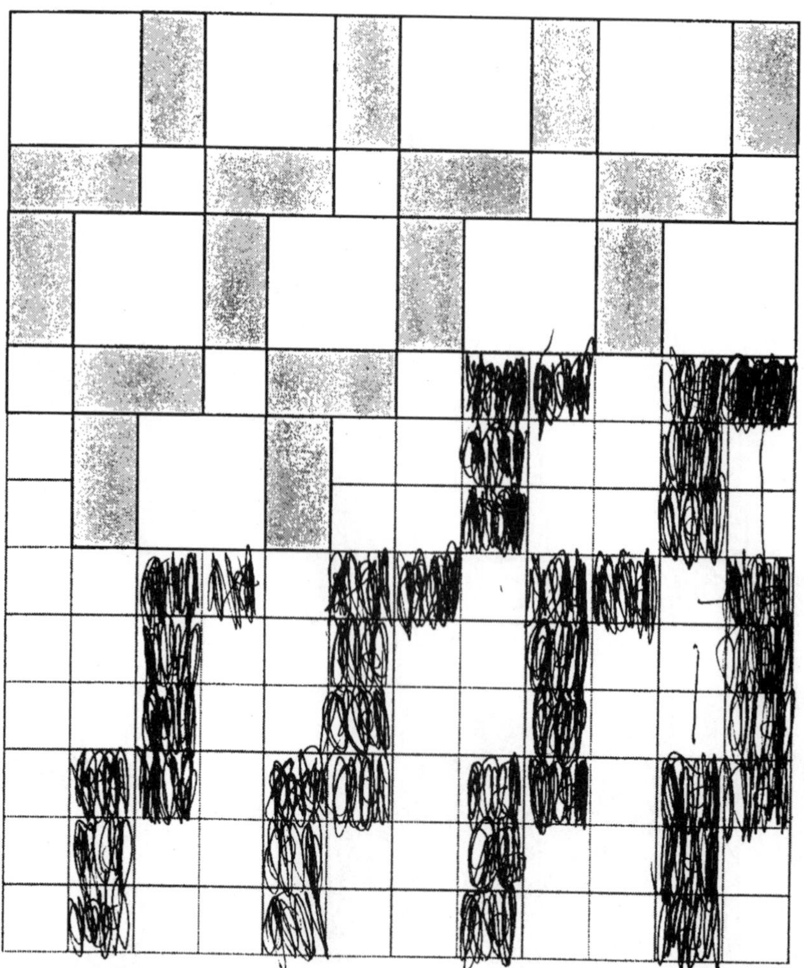

Student C

Tile Pattern

This problem gives you the chance to

- *notice and extend a geometric pattern*

Look at this tile pattern. It isn't finished yet. Finish drawing the pattern on the grid. On some parts of the edge, you may be able to show only half a tile.

Student D

Tile Pattern

This problem gives you the chance to

- *notice and extend a geometric pattern*

Look at this tile pattern. It isn't finished yet. Finish drawing the pattern on the grid. On some parts of the edge, you may be able to show only half a tile.

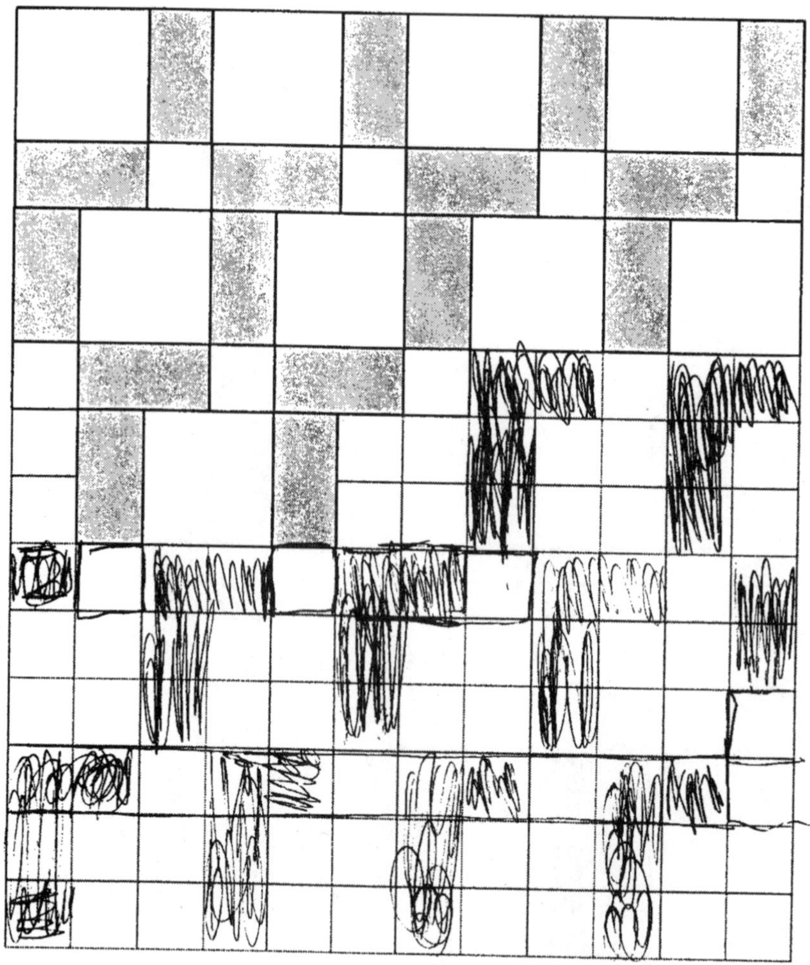

Task 16

Overview

Interpret data in a table.
Complete a pictograph.

Danville's Library

Short Task

Task Description

In this task students are asked to use information from a table to complete a pictograph.

Assumed Mathematical Background

Students should have had some experience interpreting data in a variety of forms and in creating simple graphs.

Core Elements of Performance

- interpret data in a table
- complete a pictograph

Circumstances

Grouping:	Students work to complete an individual written response.
Materials:	No special materials are needed for this task.
Estimated time:	5 minutes

Name Date

Danville's Library

This problem gives you the chance to

- *finish a pictograph accurately*

The public library in Danville asked members to vote for one new service. This is how members voted:

Audiotapes	378 votes
Foreign-Language Newspapers	199 votes
Videotapes	421 votes
CDs	755 votes

Complete the pictograph so that it shows the results as accurately as possible.

Which New Library Service?
Voting Results

Audiotapes 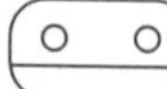 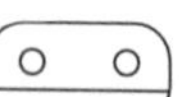  

Foreign-Language Newspapers 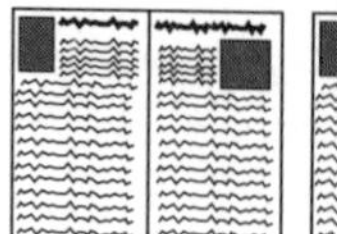 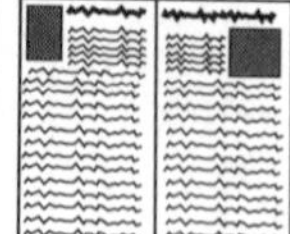

Videotapes

CDs

Each Completed Picture = 100 Votes

© The Regents of the University of California

A Sample Solution

Task

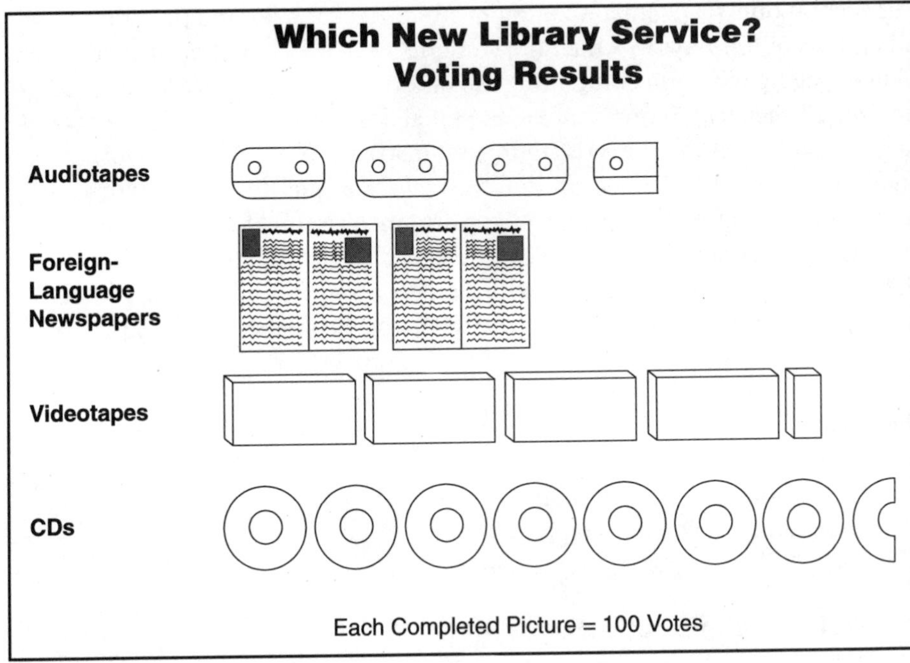

Characterizing Performance

This section offers a characterization of student responses and provides indications of the ways in which the students were successful or unsuccessful in engaging with and completing the task. The descriptions are keyed to the *Core Elements of Performance.* Our global descriptions of student work range from "The student needs significant instruction" to "The student's work meets the essential demands of the task." Samples of student work that exemplify these descriptions of performance are included below, accompanied by commentary on central aspects of each student's response. These sample responses are *representative;* they may not mirror the global description of performance in all respects, being weaker in some and stronger in others.

The characterization of student responses for this task is based on these *Core Elements of Performance:*

1. Interpret data in a table.
2. Complete a pictograph.

Descriptions of Student Work

The student needs significant instruction.

These papers show, at most, evidence of understanding the number of multiples of 100 for one of the categories (either videotapes or CDs), or how to draw an accurate partial icon for a number less than 100 for one of the categories.

Student A

This response shows that only the full icons for the videotapes are accurate (4). There is no partial icon representing 21 videotapes; there are 6 full icons for CDs, rather than 7; the partial icon for the CDs is clearly less than half an icon, which it should not be.

The student needs some instruction.

These papers provide evidence of understanding one element of the pictograph well or evidence that one of the two categories was represented well. The paper may show understanding in both the videotape and CD categories that multiples of 100 = 1 icon and make a representation that is

accurate to a degree, but only to the nearest 100 (or the next lower 100). Alternatively, the response may show that one of the two categories is represented accurately, but the other is inaccurate both for whole and partial icons.

Student B

This response shows an accurate number of full icons for each of the two categories, if the actual amounts are rounded to the nearest 100. There are 4 videotape icons and 8 CD icons with no partial icons shown.

The student's work needs to be revised.

In these papers one of the categories (videotapes or CDs) is represented well. Additionally, the second category is accurate for either the whole or the partial icons.

Typically, these papers show inaccuracy in one of the partial icons. This happens most often with the partial icon for the videotapes category, which may be misrepresented as half an icon.

Student C

This response shows an accurate number of full icons for each of the two categories. The partial icon for the CDs is a little more than half, which is accurate. There is no partial icon for the videotape category representing the 21 videotapes over 400.

The student's work meets the essential demands of the task.

These papers show that the pictograph is accurately completed. Both categories are represented well. The videotapes representation shows 4 full icons and another that is about one-fifth of an icon, or clearly less than half. The CDs category shows 7 full icons and another that is fully one half, but less than three fourths.

Student D

There is an accurate number of full icons for each of the two categories. The partial icon for the videotape category is about one fourth of an icon; the partial icon for the CD category is a little more than one half, which is accurate.

Student A

This problem gives you the chance to

- *finish a pictograph accurately*

The public library in Danville asked members to vote for one new service. This is how members voted:

Audiotapes	378 votes
Foreign-Language Newspapers	199 votes
Videotapes	421 votes
CDs	755 votes

Complete the pictograph so that it shows the results as accurately as possible.

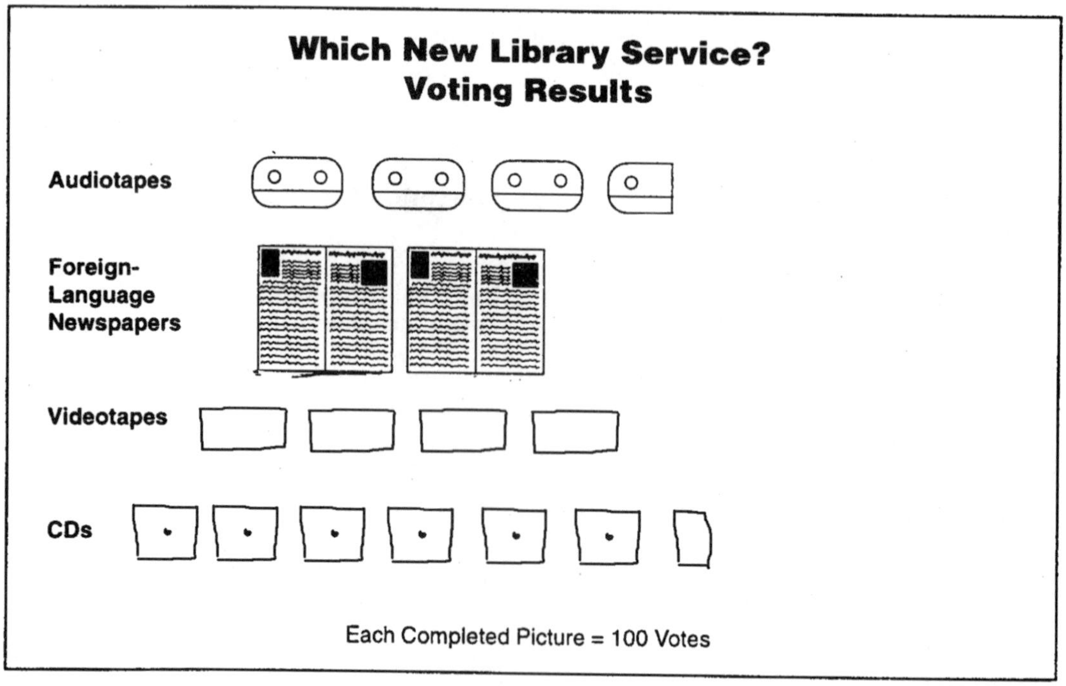

Which New Library Service?
Voting Results

Audiotapes

Foreign-Language Newspapers

Videotapes

CDs

Each Completed Picture = 100 Votes

Student B

This problem gives you the chance to

- *finish a pictograph accurately*

The public library in Danville asked members to vote for one new service. This is how members voted:

Audiotapes	378 votes
Foreign-Language Newspapers	199 votes
Videotapes	421 votes
CDs	755 votes

Complete the pictograph so that it shows the results as accurately as possible.

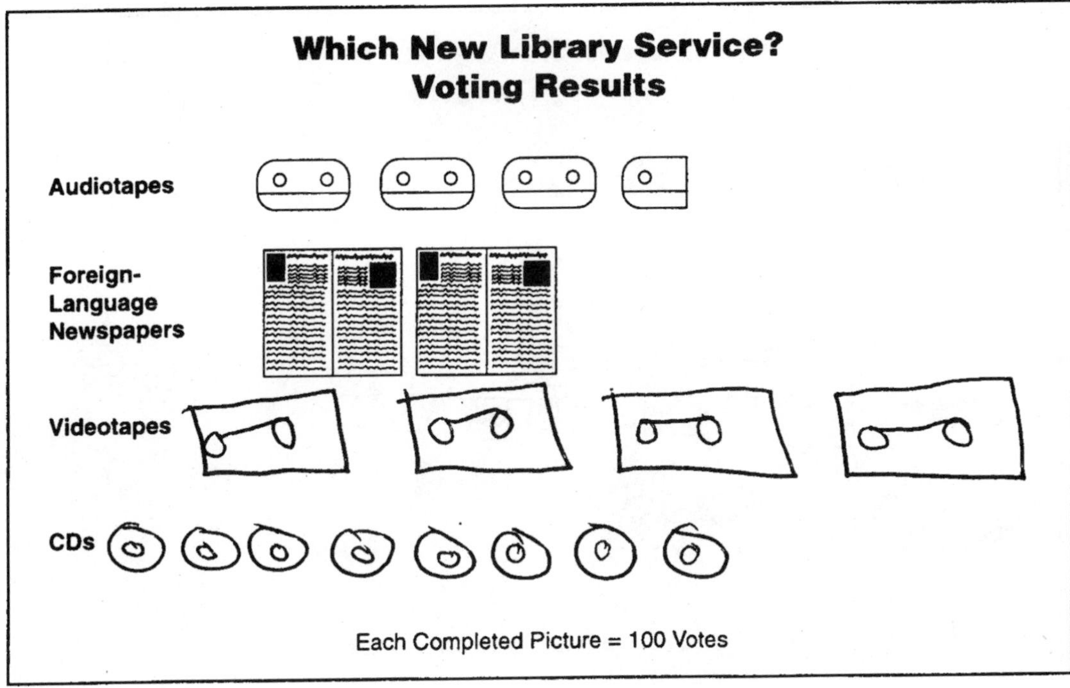

Student C

This problem gives you the chance to

- *finish a pictograph accurately*

The public library in Danville asked members to vote for one new service. This is how members voted:

Audiotapes	378 votes
Foreign-Language Newspapers	199 votes
Videotapes	421 votes
CDs	755 votes

Complete the pictograph so that it shows the results as accurately as possible.

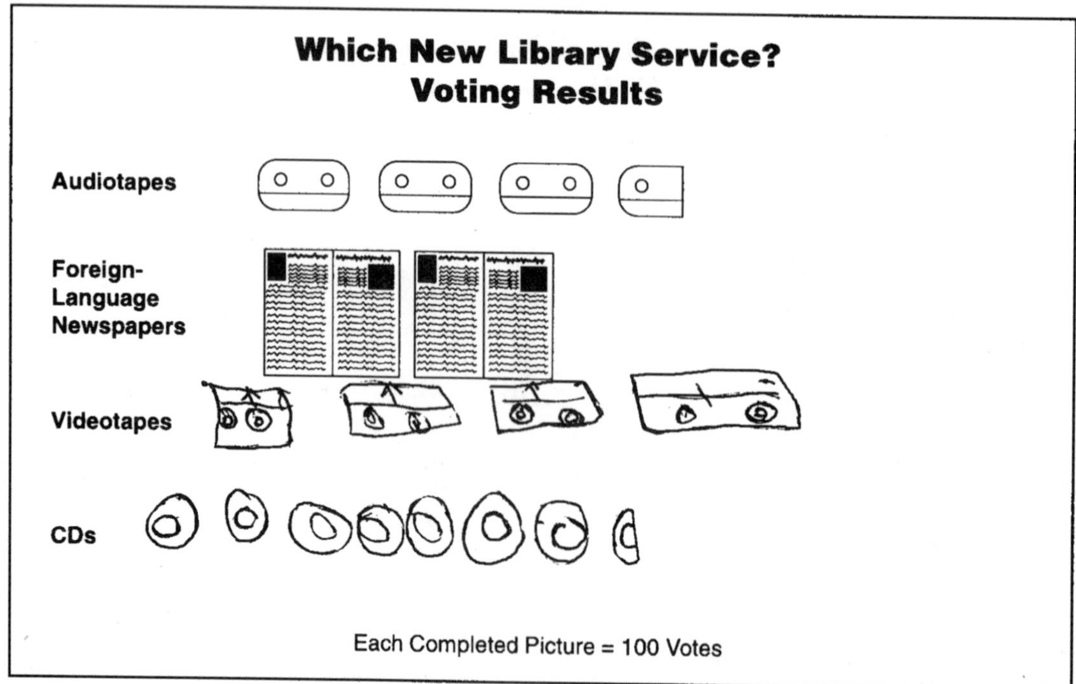

Student D

This problem gives you the chance to

- *finish a pictograph accurately*

The public library in Danville asked members to vote for one new service. This is how members voted:

Audiotapes	378 votes
Foreign-Language Newspapers	199 votes
Videotapes	421 votes
CDs	755 votes

Complete the pictograph so that it shows the results as accurately as possible.

Which New Library Service?
Voting Results

Audiotapes

Foreign-Language Newspapers

Videotapes

CDs

Each Completed Picture = 100 Votes

Glossary

This glossary defines a number of the terms that are used to describe the *Dimensions of Balance* table that appears in the package Introduction.

Applied power: a task goal—to provide students an opportunity to demonstrate their power over a real-world practical situation, with that as the main criterion for success. This includes choosing mathematical tools appropriately for the problem situation, using them effectively, and interpreting and evaluating the results in relation to the practical needs of the situation. [cf. *illustrative application*]

Checking and evaluating: a mathematical process that involves evaluating the quality of a problem solution in relation to the problem situation (for example, checking calculations; comparing model predictions with data; considering whether a solution is reasonable and appropriate; asking further questions).

Definition of concepts: a task type—such tasks require the clarification of a concept and the generation of a mathematical definition to fit a set of conditions.

Design: a task type that calls for the design, and perhaps construction, of an object (for example, a model building, a scale drawing, a game) together with instructions on how to use the object. The task may include evaluating the results in light of various constraints and desirable features. [cf. *plan*]

Evaluation and recommendation: a task type that calls for collecting and analyzing information bearing on a decision. Students review evidence and make a recommendation based on the evidence. The product is a "consultant" report for a "client."

Exercise: a task type that requires only the application of a learned procedure or a "tool kit" of techniques (for example, adding decimals; solving an equation); the product is simply an answer that is judged for accuracy.

Illustrative application of mathematics: a task goal—to provide the student an opportunity to demonstrate effective use of mathematics in a context outside mathematics. The focus is on the specific piece of mathematics, while the reality and utility of the context as a model of a practical situation are secondary. [cf. *applied power*]

Inferring and drawing conclusions: a mathematical process that involves applying derived results to the original problem situation and interpreting the results in that light.

Modeling and formulating: a mathematical process that involves taking the situation as presented in the task and formulating mathematical statements of the problem to be solved. Working the task involves selecting appropriate representations and relationships to model the problem situation.

Nonroutine problem: a task type that presents an unfamiliar problem situation, one that students are not expected to have analyzed before or have not met regularly in the curriculum. Such problems demand some flexibility of thinking, and adaptation or extension of previous knowledge. They may be situated in a context that students have not encountered in the curriculum; they may involve them in the introduction of concepts and techniques that will be explicitly taught at a later stage; they may involve the discovery of connections among mathematical ideas.

Open-ended: a task structure that requires some questions to be posed by the student. Therefore open-ended tasks often have multiple solutions and may allow for a variety of problem-solving strategies. They provide students with a wide range of possibilities for choosing and making decisions. [cf. *open-middle*]

Open investigation: an open-ended task type that invites exploration of a problem situation with the aim of discovering and establishing facts and relationships. The criteria for evaluating student performance are based on exploring thoroughly, generalizing, justifying, and explaining with clarity and economy.

Open-middle: a task structure in which the question and its answer are well-defined (there is a clear recognizable "answer") but with a variety of strategies or methods for approaching the problem. [cf. *open-ended*]

Plan: a task type that calls for the design of a sequence of activities, or a schedule of events, where time is an essential variable and where the need to organize the efforts of others is implied. [cf. *design*]

Pure mathematics: a task type—one that provides the student an opportunity to demonstrate power over a situation within a mathematics "microworld.." This may be an open investigation, a nonroutine problem, or a technical exercise.

Reporting: a mathematical process that involves communicating to a specified "audience" what has been learned about the problem. Components of a successful response include explaining why the results follow from the problem formulation, explaining manipulations of the formalism, and drawing conclusions from the information presented, with some evaluation.

Re-presentation of information: a task type that requires interpretation of information presented in one form and its translation to some different form (for example, write a set of verbal directions that would allow a listener to reproduce a given geometric design; represent the information in a piece of text with a graphic or a symbolic expression).

Review and critique: a task type that involves reflection on curriculum materials (for example, one might review a piece of student work, identify errors, and make suggestions for revision; pose further questions; produce notes on a recently learned topic).

Scaffolding: the degree of detailed step-by-step guidance that a task prompt provides a student.

Task length: the time that should be allowed for students to work on the task. Also important is the length of time students are asked by the task to think independently—the reasoning length. (For a single well-defined question, reasoning length will equal the task length; for a task consisting of many parts, the reasoning length can be much shorter—essentially the time for the longest part.)

Transforming and manipulating: a mathematical process that involves manipulating the mathematical forms in which the problem is expressed, usually with the aim of transforming them into other equivalent forms that represent "solutions" to the problem (for example, dividing one fraction by another, making a geometric construction, solving equations, plotting graphs, finding the derivative of a function).